Developed and Published
by
AIMS Education Foundation

This book contains materials developed by the AIMS Education Foundation. **AIMS** (**A**ctivities **I**ntegrating **M**athematics and **S**cience) began in 1981 with a grant from the National Science Foundation. The non-profit AIMS Education Foundation publishes hands-on instructional materials that build conceptual understanding. The foundation also sponsors a national program of professional development through which educators may gain expertise in teaching math and science.

AIMS Education Foundation
P.O. Box 8120, Fresno, CA 93747-8120 • 888.733.2467 • aimsedu.org

ISBN 978-1-932093-98-8

Printed in the United States of America

I Hear and I Forget,
I See and I Remember,
I Do and I Understand.

-Chinese Proverb

Table of Contents

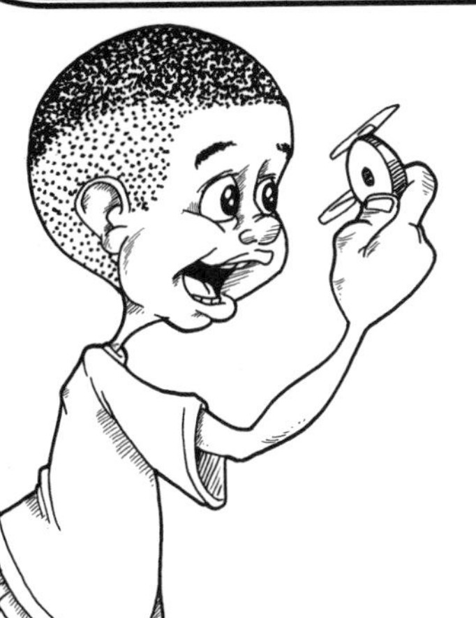

In this book, students will explore the invisible force of magnetism. Through hands on experiences, they will discover the following concepts:

Every magnet has a magnetic field that interacts with the magnetic fields of objects containing iron or other magnetic materials. Most magnetic objects students will use and be acquainted with are made of some form of iron.

Magnets usually have two poles, a north-seeking and a south-seeking pole.

When two magnets are placed near one another, they react according to the poles that are near one another. Unlike poles attract, and like poles repel.

Magnets can attract magnetic materials through all non-magnetic and most magnetic materials.

Magnets vary in strength.

Since magnetic force is greater than that of gravity, magnetism can be used to defy gravity in various ways.

Magnetism is a non-contact force. It interacts with objects at a distance.

SCIENCE INFORMATION

Historical Background

The word magnet is derived from Magnesia, the name of a region of Greece where certain dark, metallic rocks called *lodestones* were found in ancient times. Lodestone is really a kind of *magnetite*, a hard rock with high iron content and magnetic properties. People in Europe and China discovered that if a lodestone were placed on a piece of wood and then floated in a bowl of water, the same part of the lodestone would always point toward the Polestar or North Star. This earliest *magnetic compass*, also called a *directional* or *navigational* compass, was welcomed by mariners, especially for use on long voyages when stars and familiar shore landmarks were not available to guide them.

As with other of the world's major scientific discoveries, many changes have been made to improve the usefulness of the magnetic compass. A magnetized needle replaced the lodestone. By the thirteenth century, the needle was placed on a pivot, so it could move more freely. A number of changes in the compass and its base enabled mariners to read it, even when seas were rough. The discovery of the magnetic north pole and the building of iron ships prompted still other refinements. At first, only the cardinal compass points (north, south, west, east) were identified on compasses. The desire for more precise markings led to the identification of the degrees of a circle. Throughout the history of the compass, the basic principle of the magnetic compass has remained the same. Even in this computer age, it is still numbered among navigational tools considered essential.

Magnetic Poles

The Earth acts as if there were a large bar magnet running from north to south through the middle of it. The *magnetic north pole* is not at the Earth's geographic north pole; it moves constantly and has even traded places with the south magnetic pole a number of times. The most recent survey done by the Geological Survey of Canada (May, 2001) locates the north magnetic pole in the Canadian Arctic, Latitude 81.3 degrees North and 110.8 degrees West. This location lies near Ellesmere Island in northern Canada. They have established that the pole's location moves northwest approximately 40 km (25 mi) per year.

Individual magnets also have magnetic poles with unique but consistent characteristics. Magnetic force is always strongest at the poles. Normal magnets have two poles; in industry or research they may have more but usually an even number. There can never be a magnet with only one pole. If two magnets are placed facing each other with like magnetic poles together, they will repel one another; if unlike poles are together, they will attract. If a magnet is broken, each piece becomes another complete magnet with the correct, complementary pole forming at each of the broken ends.

Magnetic Domains

Why do magnets and magnetic materials act the way they do? The answer lies in the fact that iron and other magnetic materials are made up of very small areas called *magnetic domains*. Each domain consists of billions of atoms. The atoms within each individual domain are always perfectly aligned magnetically. In contrast, the domains themselves may or may not be aligned.

Physicists are constantly learning more about *domain alignment*, and even the definition of the terminology has changed. Until recently it was thought that when domains become aligned, they change the direction in which they point. Current research indicates, however, that in the process of domain alignment, the domains pointing in the right direction probably become larger at the expense of those pointing in other directions.

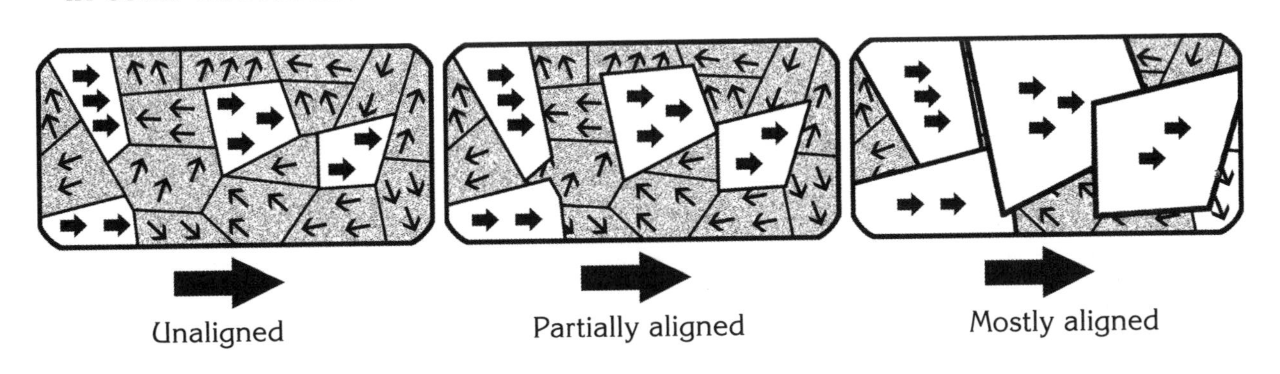

The proportion of domains that are aligned determines the amount of magnetic force a magnet has. A piece of pure iron might have almost all unaligned domains and very little magnetic strength. If a moderately strong magnet is held near this iron or strokes it, some of the iron's domains will become aligned and it will exhibit some magnetic properties.

How do the iron's domains know which way to align themselves? The answer depends on the pole of the magnet used to touch or stroke the iron. Since opposites attract, touching or stroking the iron with the south-seeking pole of a magnet will make that end of the iron the north-seeking pole, and the domains facing in a north-seeking direction will be the ones to grow. If only a few domains become aligned, the object will become slightly magnetic; if most of the domains become aligned, the material will be much more strongly magnetic. No material ever has all of its domains aligned.

Magnetic Fields

A *magnetic field* exists in the magnet and in the space around it. Magnetic fields are strongest near the north-seeking and south-seeking poles. The shape of a magnet's magnetic field can be shown by using iron bits or filings to reveal the *magnetic field lines* that spread out in all directions from the north-seeking pole and back in extended arcs to the south-seeking pole. The magnetic field exists in the whole area: that is, in the spaces between the lines as well as on the lines themselves.

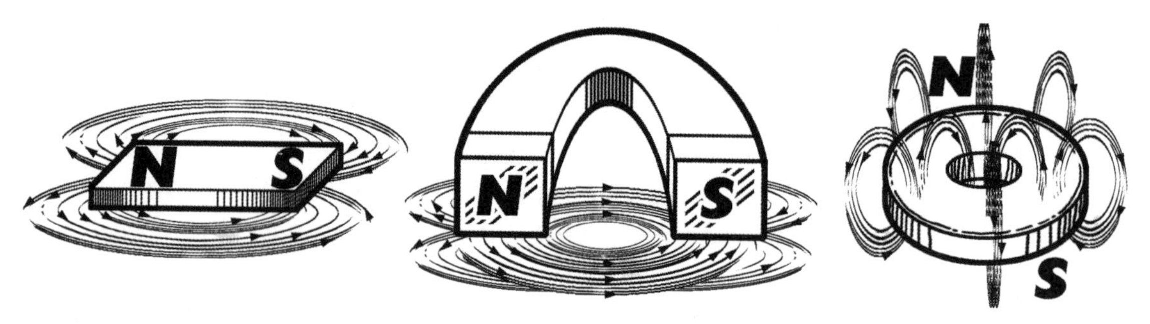

Magnetic fields are able to pass through nonmagnetic materials. This can be demonstrated by placing a piece of paper (a nonmagnetic material) between a magnet and a small magnetic object such as a paper clip; the magnetic interaction through the paper should be readily observable. If, however, the distance between the magnet and the clip is increased, the interaction may become too weak to be observed, even though the magnetic fields are still present. The effect of such nonmagnetic materials on the magnetic interaction is negligible; the main factor is the distance involved.

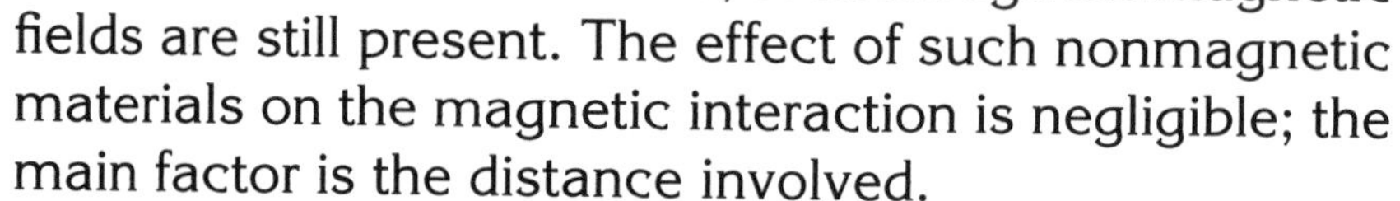

All substances actually display magnetic properties, but most show them to such a very small degree that we usually consider these materials *nonmagnetic*. Highly sophisticated scientific equipment is needed to detect magnetic characteristics at such low levels. On the other hand, a few metallic elements display magnetic properties strongly enough to be considered magnetic, or—more properly—*ferromagnetic* (*ferro* means iron). These include iron, nickel, cobalt, rare earths, plus some of their alloys like steel and strontium ferrite. All metals are not considered magnetic, a common misconception easily corrected by observing a common magnet's effect on brass, copper, or aluminum.

Magnetism as an Interaction

Magnetism is usually defined and demonstrated by focusing on an individual magnet and its magnetic field. This promotes a popular misconception that magnetic action centers on each magnet. The reality, quite the opposite, is represented by the term *magnetic interaction*. There must always be an interaction between magnetic fields, either between two magnets or between a magnet and a material such as soft iron in which a magnetic field can be induced. In other words, a magnet does not ever just lie dormant by itself somewhere; even when no one is using it to attract or repel things, its magnetic field is always interacting with other magnetic fields. Although this interaction is often not observable, it is always happening.

Attraction and repulsion would seem to be two types of magnetic force, but such is not the case. They are really two manifestations of a single magnetic force. A magnet can only attract (not repel) magnetic materials unless those materials have been magnetized. Two magnets attract or repel each other because their magnetic fields are in a certain orientation, a specific spatial position, with either like or unlike poles together.

In this book, a conscious effort has been made to use familiar vocabulary. As a result, magnets are referred to in this book as *attracting* or *repelling* materials, and objects are said to *stick to* magnets. These are not contradictions of the magnetic interaction concept, but attempts to help students build on terms most of them already use.

It is appropriate to think of magnetism this way:

- Magnetic fields interact with each other.
- A magnet attracts magnetic materials. (It can't repel them.)
- Magnets attract or repel each other.

A Word of Caution

Since magnets are essential to the functioning of some appliances, it is possible to cause damage to them by holding a magnet too close. Magnets can also erase video tapes, audio tapes, and credit cards. They can damage TV screens and computer monitors. Although we do not want to overburden students with negative restrictions that could dampen their enthusiasm, it is important to communicate some sensible safety precautions.

National Education Reform Documents

The AIMS Education Foundation is committed to remaining at the cutting edge of providing curricular materials that are user-friendly, educationally sound, developmentally appropriate, and aligned with the recommendations found in national education reform documents.

Project 2061 Benchmarks

The Nature of Science

- *People can often learn about things around them by just observing those things carefully, but sometimes they can learn more by doing something to the things and noting what happens.*

The Physical Setting

- *Objects can be described in terms of the materials they are made of (clay, cloth, paper, etc.) and their physical properties (color, size, shape, weight, texture, flexibility, etc.).*
- *Things move in many different ways, such as straight, zigzag, round and round, back and forth, and fast and slow.*
- *Magnets can be used to make some things move without being touched.*
- *Without touching them, a magnet pulls on all things made of iron and either pushes or pulls on other magnets.*

The Mathematical World

- *Often a person can find out about a group of things by studying just a few of them.*

Habits of Mind

- *Raise questions about the world around them and be willing to seek answers to some of them by making careful observations and trying things out.*
- *Keep records of their investigations and observations.*

American Association for the Advancement of Science.
Benchmarks for Science Literacy
Oxford University Press. New York. 1993

National Science Education Standards

Abilities Necessary to do Scientific Inquiry

- *Plan and conduct a simple investigation.*
- *Employ simple equipment and tools to gather data and extend the senses.*
- *Communicate investigations and explanations.*

Properties of Objects and Materials

- *Objects are made of one or more materials, such as paper, wood, and metal. Objects can be described by the properties of the materials from which they are made, and those properties can be used to separate or sort a group of objects or materials.*

Position and Motion of Objects

- *An object's motion can be described by tracing and measuring its position over time.*
- *The position and motion of objects can be changed by pushing or pulling. The size of the change is related to the strength of the push or pull.*

Light, Heat, Electricity, and Magnetism

- *Magnets attract and repel each other and certain kinds of other materials.*

National Research Council
National Science Education Standards
National Academy Press. Washington, DC. 1996

NCTM Standards 2000*

Number and Operations

- *Develop and use strategies for whole-number computations, with a focus on addition and subtraction*
- *Understand the effects of adding and subtracting whole numbers*

Algebra

- *Sort, classify, and order objects by size, number, and other properties*

Data Analysis and Probability

- *Sort and classify objects according to their attributes and organize data about the objects*
- *Represent data using concrete objects, pictures, and graphs*

Reprinted with permission from
Principles and Standards for School Mathematics
2000 by the National Council of Teachers of Mathematics

Topic
Magnetism

Key Question
To what will a magnet stick?

Learning Goals
Students will:
- test everyday objects to determine if they are or are not attracted to a magnet,
- sort and graph objects based on whether or not they are magnetic, and
- draw conclusions about what magnetic materials have in common.

Guiding Documents
Project 2061 Benchmarks
- *Without touching them, a magnet pulls on all things made of iron and either pushes or pulls on other magnets.*
- *Keep records of their investigations and observations.*

NRC Standards
- *Plan and conduct a simple investigation.*
- *Magnets attract and repel each other and certain kinds of other materials.*

Math
Graphing

Science
Physical science
force
magnetism

Integrated Processes
Observing
Comparing and contrasting
Classifying
Predicting
Recording

Materials
Ring magnets (see *Management 1)*
A variety of classroom objects (see *Management 2)*
Student recording page
Class graph
Masking tape
Prediction cards (see *Management 4)*

Background Information
Magnetism is an unseen force that interacts with other objects. All substances actually display magnetic properties, but most show them to such a small degree that we usually consider these materials non-magnetic. Highly sophisticated scientific equipment is needed to detect magnetic characteristics at these low levels. On the other hand, a few metallic elements such as iron, nickel, cobalt, rare earths, plus some of their alloys like steel and strontium ferrite display magnetic properties strongly enough to be considered magnetic. All metals are not considered magnetic, a common misconception easily corrected by observing a common magnet's effect on brass, copper, or aluminum.

The earliest records of magnets come from Magnesia in Asia Minor. It is believed that the word *magnet* comes from the name of the city. Lodestones, or naturally occurring magnets, were found in large amounts in the area. These natural magnets attracted pieces of iron. The modern name for lodestone is magnetite.

Safety note: Do not use magnets near electronic equipment and cards with magnetic strips, such as credit cards. Magnets will damage these objects. Also, tell children that magnets can lose their magnetic force if they are dropped.

Management
1. Ring magnets (order number 1971) are available from AIMS.
2. Prior to teaching this lesson, gather objects to test, such as keys, coins, paper clips, brass paper fasteners, scissors, erasers, pencils, safety pins, soda cans, etc.

3. Prepare a concrete class graph by placing *Stick* and *No Stick* labels on the floor and dividing the two categories with a strip of masking tape.
4. Copy a set of *Stick* and *No Stick* prediction cards for each student.

Procedure

1. Ask the *Key Question*. Record student responses on the board.
2. Give each child a set of *Stick* and *No Stick* prediction cards. Practice using the cards so the children can tell which card shows that the object stuck to the magnet and which shows that the item did not stick to the magnet.
3. Choose an object. Discuss the characteristics of the object, including what it is made of—plastic, metal, paper, wood, etc. Have students use their prediction cards to indicate whether they think a magnet will stick to the object or not.
4. Invite a student to touch the object with a magnet and determine if the magnet sticks or not and then place the object in the appropriate column of the class graph.
5. Repeat procedures three and four with several additional objects.
6. When all of the objects have been explored, discuss the results as shown by the graph. Generalize what type of materials a magnet will stick to.
7. When students are familiar with the process, divide them into small groups and allow them to choose four objects from the classroom to sort based on their interaction with a magnet—stick or no stick.
8. Distribute the student page and have students record the results of the exploration in either pictures or words.
9. End with a time for students to share their results.

Connecting Learning

1. How did we sort our objects?
2. What kind of objects stuck to the magnet? What kinds didn't stick?
3. Were your predictions about the different objects always correct? Explain.
4. What other objects did you test? What were the results?
5. Do all metal things stick to a magnet? Explain your thinking.
6. Did anyone find an object that one part of the object stuck to the magnet and another part did not? Explain.

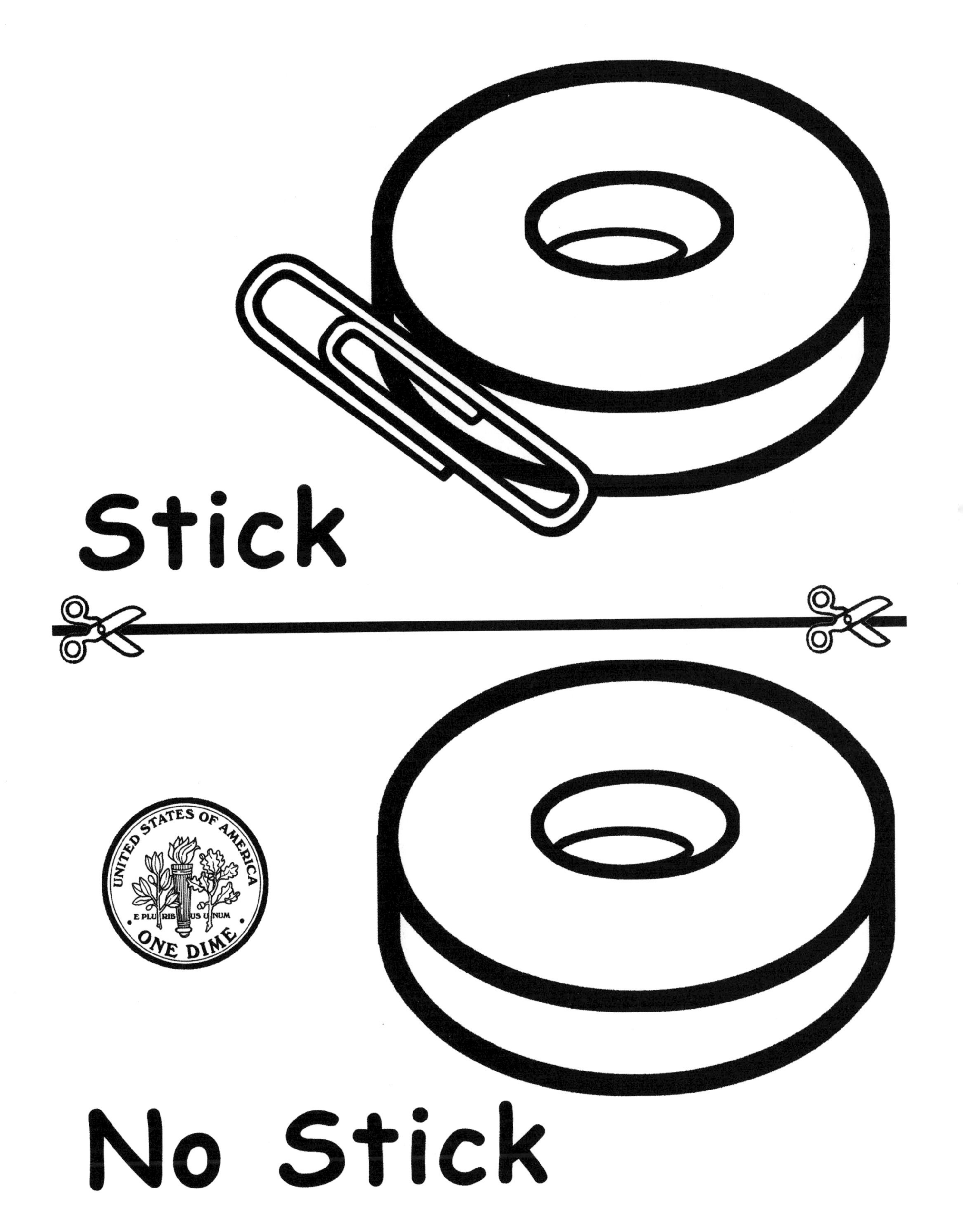
Stick
UNITED STATES OF AMERICA
ONE DIME
No Stick

My magnet will stick to these.	My magnet will not stick to these.

Topic
Magnetism

Key Question
To what types of household cans is a magnet attracted?

Learning Goal
Students will sort cans based on their attraction to a magnet.

Guiding Documents
Project 2061 Benchmarks
- *Without touching them, a magnet pulls on all things made of iron and either pushes or pulls on other magnets.*
- *Keep records of their investigations and observations.*

NRC Standards
- *Plan and conduct a simple investigation.*
- *Magnets attract and repel each other and certain kinds of other materials.*
- *The position and motion of objects can be changed by pushing or pulling. The size of the change is related to the strength of the push or pull.*

*NCTM Standards 2000**
- *Sort, classify, and order objects by size, number, and other properties*
- *Sort and classify objects according to their attributes and organize data about the objects*

Math
Sorting

Science
Physical science
  force
    magnetism

Integrated Processes
Observing
Comparing and contrasting
Classifying
Predicting
Recording
Generalizing
Applying

Materials
Ring magnets (see *Management 2)*
Cans (see *Management 1)*
Student page

Background Information
Magnets attract magnetic objects. Magnetic objects are generally metal—most commonly made of iron, cobalt, and/or nickel. Not all metal objects are magnetic. Aluminum, of which most soda cans are made, is not magnetic.

In this activity, students will sort common household cans based on their attraction to a magnet. Through this, they will discover that all metals are not attracted to a magnet and can begin to generalize which metals are and which are not.

Management
1. Each group will need several cans made of different metals, both magnetic and non-magnetic. Suggested cans would include soda, tuna, cat food, soup, coffee, nut, potato chip, etc. The cans need not be the same for each group.
2. Ring magnets are available from AIMS (order number 1971). Each group needs a magnet.

Procedure
1. Display several cans made of both metallic and nonmetallic materials for the class to see. Review the term *attract* as it relates to magnets. Ask the class if they think all of the cans would be attracted to a magnet. Discuss their thinking.
2. Divide the class into groups two or three.

3. Give each group a set of cans, a student page, and a magnet.
4. Instruct the groups to predict, test, and record results using pictures and/or words. Provide time for students to sort their cans based on attraction to the magnet.
5. When all groups have completed their testing, bring the class together to discuss their results.
6. Have each group bring their magnetic cans up and place them together; then have them bring their nonmagnetic cans and place them together. Compare the two groups of cans and allow students to generalize what types of metals are magnetic and which are not.

Connecting Learning
1. What does the word *attract* mean?
2. Are all cans attracted to a magnet? Why or why not? [No. Not all cans are made of magnetic materials.]
3. What types of cans were attracted to your magnet? [cans containing iron] Why? [Iron is a magnetic material.]
4. Which type of can did you have more of—those that were attracted to a magnet or those that were not attracted? How did you know there were more? How many more were there?
5. Would a can of corn be magnetic? Why or why not?
6. What are you wondering now?

* Reprinted with permission from *Principles and Standards for School Mathematics*, 2000 by the National Council of Teachers of Mathematics. All rights reserved.

Attracted (magnetic) | Not Attracted (non-magnetic)

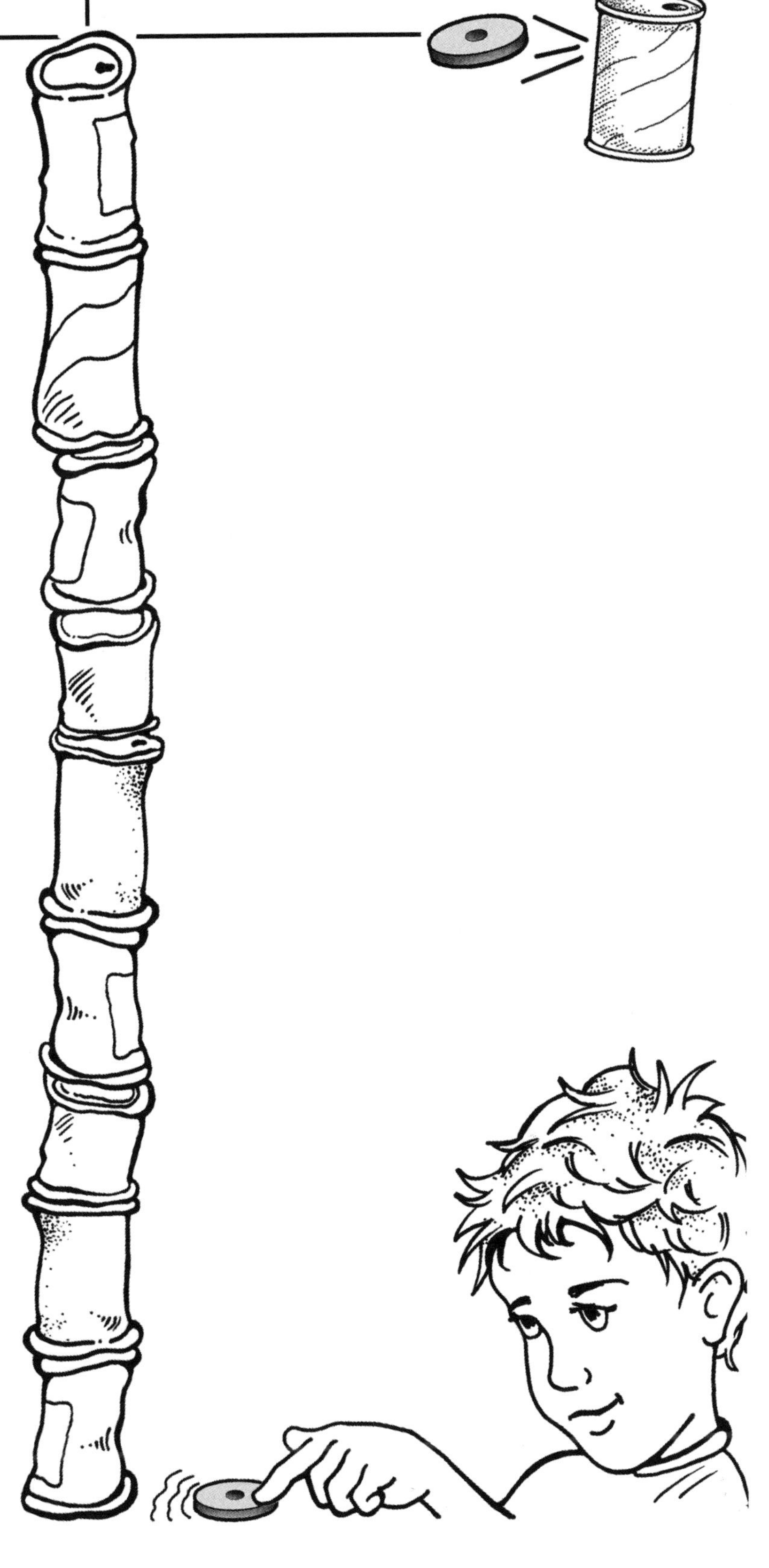

Topic
Magnetism

Key Question
What kinds of objects are magnetic?

Learning Goals
Students will:
- test everyday objects to see which are attracted to a magnet, and
- earn money for correctly identifying items that are magnetic.

Guiding Documents
Project 2061 Benchmark
- *Without touching them, a magnet pulls on all things made of iron and either pushes or pulls on other magnets.*

NRC Standard
- *Magnets attract and repel each other and certain kinds of other materials.*

*NCTM Standards 2000**
- *Develop and use strategies for whole-number computations, with a focus on addition and subtraction*
- *Understand the effects of adding and subtracting whole numbers*

Math
Addition
Money

Science
Physical science
force
magnetism

Integrated Processes
Observing
Comparing and contrasting
Predicting
Recording

Materials
Magnets
50 cm of string
A variety of classroom objects (see *Management 1)*
Price tags
Yardstick

Background Information
Magnets have an invisible force that interacts by attracting magnetic objects. Items made of iron, cobalt, and nickel have magnetic properties. It is important for students to understand that not all metals are attracted to a magnet. This activity will provide an opportunity for students to see that everything that is shiny is not metal and everything that is metal is not magnetic.

Safety note: Do not use magnets near electronic equipment and cards with magnetic strips, such as credit cards. Magnets will damage these objects. Also, tell children that magnets can lose their magnetic force if they are dropped.

Management
1. Prior to teaching this lesson, gather a variety of small magnetic and non-magnetic objects such as keys, coins, paper clips, brass paper fasteners, scissors, erasers, pencils, safety pins, soda cans, etc.
2. Copy the price tag page onto card stock and laminate for extended use. Attach a price tag to each of the objects you collected. It is suggested that you attach a large value to an item that is non-magnetic, such as an aluminum soda can, to see if students will be drawn to the amount or to what they think is magnetic.
3. Prepare a fishing pole-type apparatus by attaching the magnet to the string and the string to the yardstick.
4. It is suggested that students have some prior knowledge of magnets and the idea of things being magnetic and non-magnetic.
5. Ring magnets (order number 1971) are available from AIMS.

Procedure

1. Introduce the term *magnetic*.
2. Ask the *Key Question*. Record student responses on the board.
3. Tell the class that they are going to play a game that will require them to use their knowledge about magnets.
4. Divide the class into teams. Display the items you collected and point out the price tags attached to each. Explain that the price tag shows the amount of money the team will collect if they get that item. The goal is to have the most money at the end of the game. Spread the items out on a table so that they are as far apart as space will allow. Identify the table as the "pond."
5. Show students the fishing pole and describe how the game will proceed. One player from the first team will come to the pond and identify the item he/she will try to pick up using the magnet on the fishing pole. He/she will then attempt to pick up that item. If the item is magnetic and it is successfully picked up, the team takes the item, and the amount of money on the price tag is given to that team. If the item is not magnetic, it is removed from the table and no money is earned. Play continues (with a different player from each team fishing every turn) until all magnetic objects have been picked up.
6. Have the students play the game and keep track of the "funds" collected by each team on the board. At the end of the game, have the students add the money collected to determine which team had the most money.
7. End by having the teams bring all of the items they collected and placing them on one end of the table. On the other end of the table, place all the non-magnetic items. Compare and contrast the items in the two piles. Discuss what items may have tricked them, what their strategy was, etc.

Connecting Learning

1. What kinds of objects are magnetic?
2. Did your team always choose an object that was magnetic? How?
3. Did anyone find an object that had one part that was magnetic and another part that was not? Explain.
4. Are all metal objects magnetic? How do you know?
5. What are you wondering now?

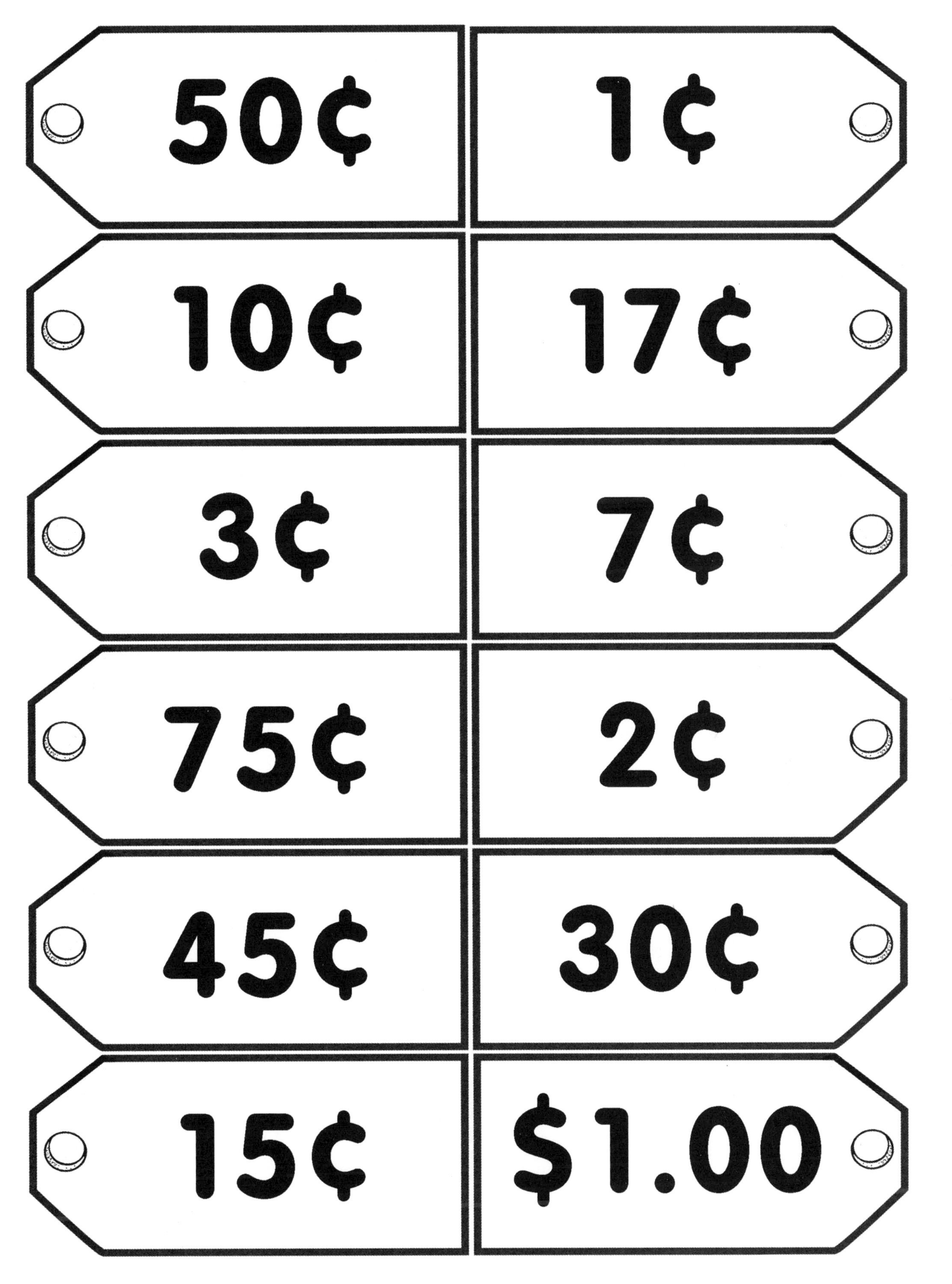
50¢
1¢
10¢
17¢
3¢
7¢
75¢
2¢
45¢
30¢
15¢
$1.00

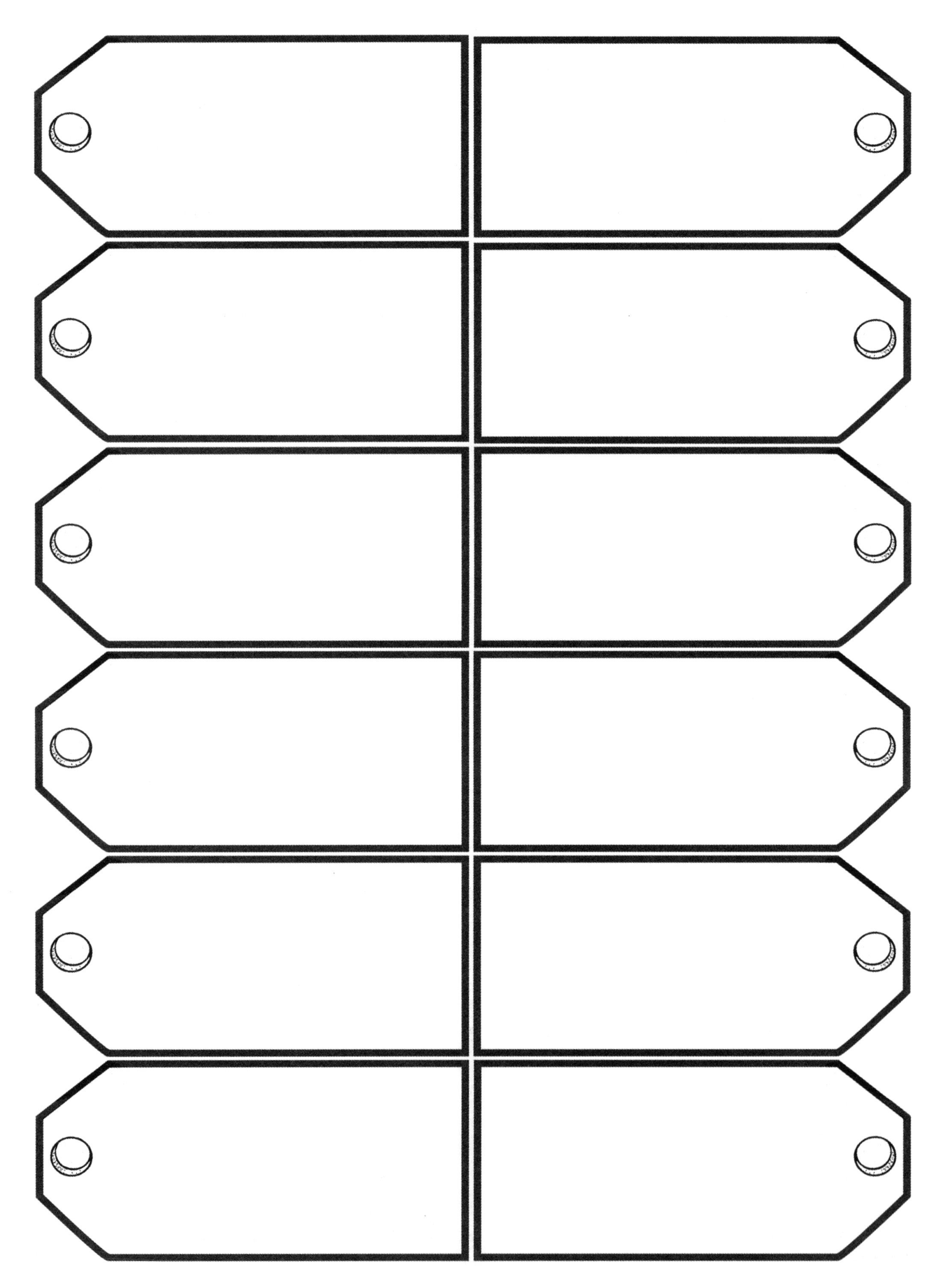

Team 1	Team 2
+	+
Total ______	Total ______

Topic
Magnets, force at a distance

Key Question
What can you find in the discovery bottle using a magnet?

Learning Goals
Students will:
- use magnets to explore force at a distance, and
- discover materials that are attracted by a magnet.

Guiding Documents
Project 2061 Benchmarks
- *Magnets can be used to make some things move without being touched.*
- *Often a person can find out about a group of things by studying just a few of them.*
- *Raise questions about the world around them and be willing to seek answers to some of them by making careful observations and trying things out.*

NRC Standards
- *Employ simple equipment and tools to gather data and extend the senses.*
- *Communicate investigations and explanations.*
- *Magnets attract and repel each other and certain kinds of other materials.*

*NCTM Standards 2000**
- *Sort, classify, and order objects by size, number, and other properties*
- *Sort and classify objects according to their attributes and organize data about the objects*

Science
Physical science
force
magnetism

Integrated Processes
Observing
Comparing and contrasting
Collecting and recording data
Communicating
Interpreting data

Materials
Clear plastic two-liter bottle
Colored gift package shreds
10-15 small objects, some that are attracted to a magnet and some that are not (see *Management 1)*
Magnets or magnet wands
Student recording sheet

Background Information
Magnets are great in a discovery center because they stimulate excitement and interest. Magnetic forces can close doors, drive toys, hold paper to refrigerators, pick up scrap steel, separate mixtures, and even move trains. Young children need many, and repeated, opportunities to observe the attractive force of magnets, a simple yet very complex phenomenon. Students can observe that magnets can be used to make some things move, while others do not. They notice that the magnet can make things move without touching them. The magnet may pull another object and push or pull another magnet.

Through experimentation, students can begin to understand that some things happen even though the force that causes those things to happen cannot be observed. The conversation generated by the discovery bottle is fodder for teacher observation and future investigations.

Management
1. Prepare the two-liter bottle by filling the bottle with the packaging shreds and then dropping in the small objects that are and are not attracted to a magnet (for example: paper clip, paper fastener, marble, counters, keys, etc.).
2. Students should be grouped in pairs or groups of four. This activity may be carried over for several days to allow time for all students to visit the discovery center.
3. Copy a recording sheet for each student.

Procedure
Part One
1. Ask the *Key Question.*
2. Introduce the discovery bottle and model for students using the magnet how to explore the bottle.
3. Invite several students to come, one at a time, and run the magnet over the outside of the bottle and describe what they see.

4. Model by drawing a picture or writing the words on the student recording sheet to show which objects were attracted by the magnet.

Part Two
1. During center or discovery time, have students explore the discovery bottle and draw pictures of objects they can find in the bottle while using the magnet.
2. Gather the students together to share their discoveries. List the objects on the board in two columns, *Items attracted to a magnet* and *Items not attracted to a magnet.*
3. Encourage students to compare and discuss the objects that they discovered.

Connecting Learning
1. What objects did you find in the discovery bottle?
2. Did all of us see the same things? Why?
3. How did the magnet help you find objects in the bottle?
4. What kinds of objects did you find?
5. Did you see any objects that were not attracted to the magnet? Why do you think they were not attracted?
6. Were there any objects that were attracted to the magnet that surprised you? What were they and why were you surprised?
7. Can you think of some other objects we could put into the discovery bottle?
8. Did your magnet have to touch the object in order to attract it? Explain.
9. What are some of the ways that you use magnets at home?

* Reprinted with permission from *Principles and Standards for School Mathematics,* 2000 by the National Council of Teachers of Mathematics. All rights reserved.

Discovery Bottle
This is what I discovered...

Topic
Magnetism

Key Question
To what will a magnet stick?

Learning Goal
Students will play a game that requires them to correctly predict whether or not objects are magnetic to win.

Guiding Documents
Project 2061 Benchmark
- *Without touching them, a magnet pulls on all things made of iron and either pushes or pulls on other magnets.*

NRC Standard
- *Magnets attract and repel each other and certain kinds of other materials.*

Science
Physical science
  force
    magnetism

Integrated Processes
Observing
Comparing and contrasting
Classifying
Predicting

Materials
For each group of students:
  die
  set of game cards
  game board
  objects (see *Management 2)*
  ring magnet
  game pieces

Background Information
Magnets have an invisible force that attracts objects made of iron, cobalt, and nickel. These are called magnetic materials.

It is important for students to understand that all metals are not magnetic. This activity will provide an opportunity for students to see that everything that is shiny is not metal and everything that is metal is not magnetic.

Management
1. Ring magnets (order number 1971) are available from AIMS.
2. Prior to teaching this lesson, gather a set of the objects pictured on the game cards for each group of students. Place them in re-closable bags for ease in distribution.
3. Copy a set of the game cards on card stock and laminate for extended use. Each group needs a set of cards.
4. This activity can serve as an assessment or as an opportunity to explore magnetic and non-magnetic materials. If it is used as exploration, the terms *magnetic* and *non-magnetic* will need to be clarified.

Procedure
1. Review the terms *magnetic* and *non-magnetic*. Ask students to name objects that they think are magnetic.
2. Explain to students that they will be getting into small groups and will play a game where they will predict whether an object is magnetic or not, then they will test to see if they were correct.
3. Put the class into small groups and give each group of students a game piece for each member, a game board, a die, a magnet, a set of the game cards, and a bag of objects that match the cards.
4. Have the students roll the dice and move their game pieces the correct number of spaces. If they land on a magnet space, they draw a card and predict whether the object on the card is magnetic or not. After they have made their prediction, they are to take a magnet and try to pick up the object pictured on the card to see if their prediction was correct. If they were correct, they get to hold on to the card they chose and roll the dice again on their next turn. If they were incorrect, they are to put their card at the bottom of the card pile. On their next turn they will not roll the dice, they just choose another card. The first to the finish line wins.

Connecting Learning

1. What did you learn from the game?
2. What was your favorite part of the game?
3. Do you think it would be important for other students that are learning about magnets to play this game? Why?
4. What other magnetic objects could we have used? What non-magnetic objects would you include?
5. What are you wondering now?

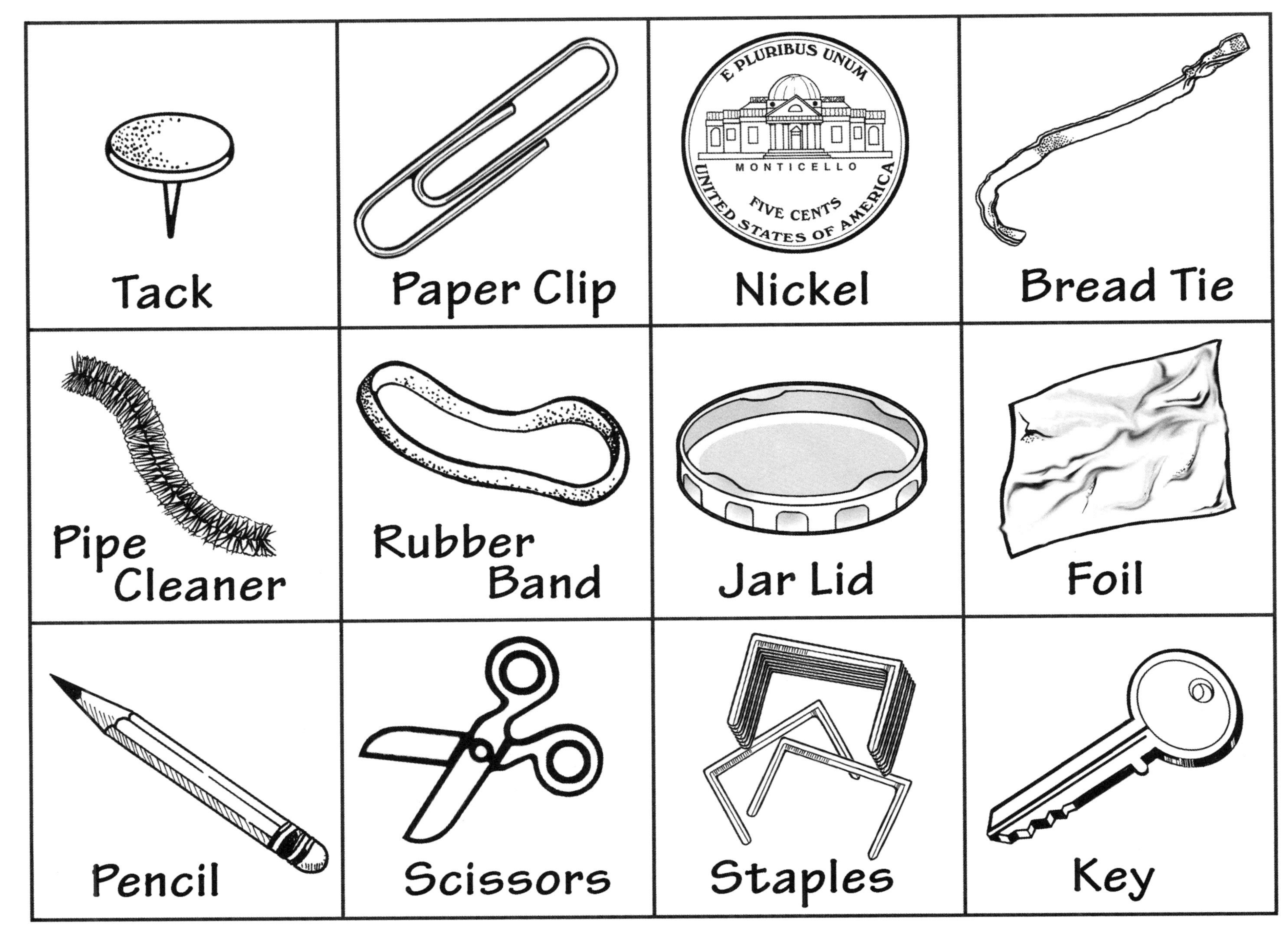

Tack
Paper Clip
E PLURIBUS UNUM
MONTICELLO
FIVE CENTS
UNITED STATES OF AMERICA
Nickel
Bread Tie
Pipe Cleaner
Rubber Band
Jar Lid
Foil
Pencil
Scissors
Staples
Key

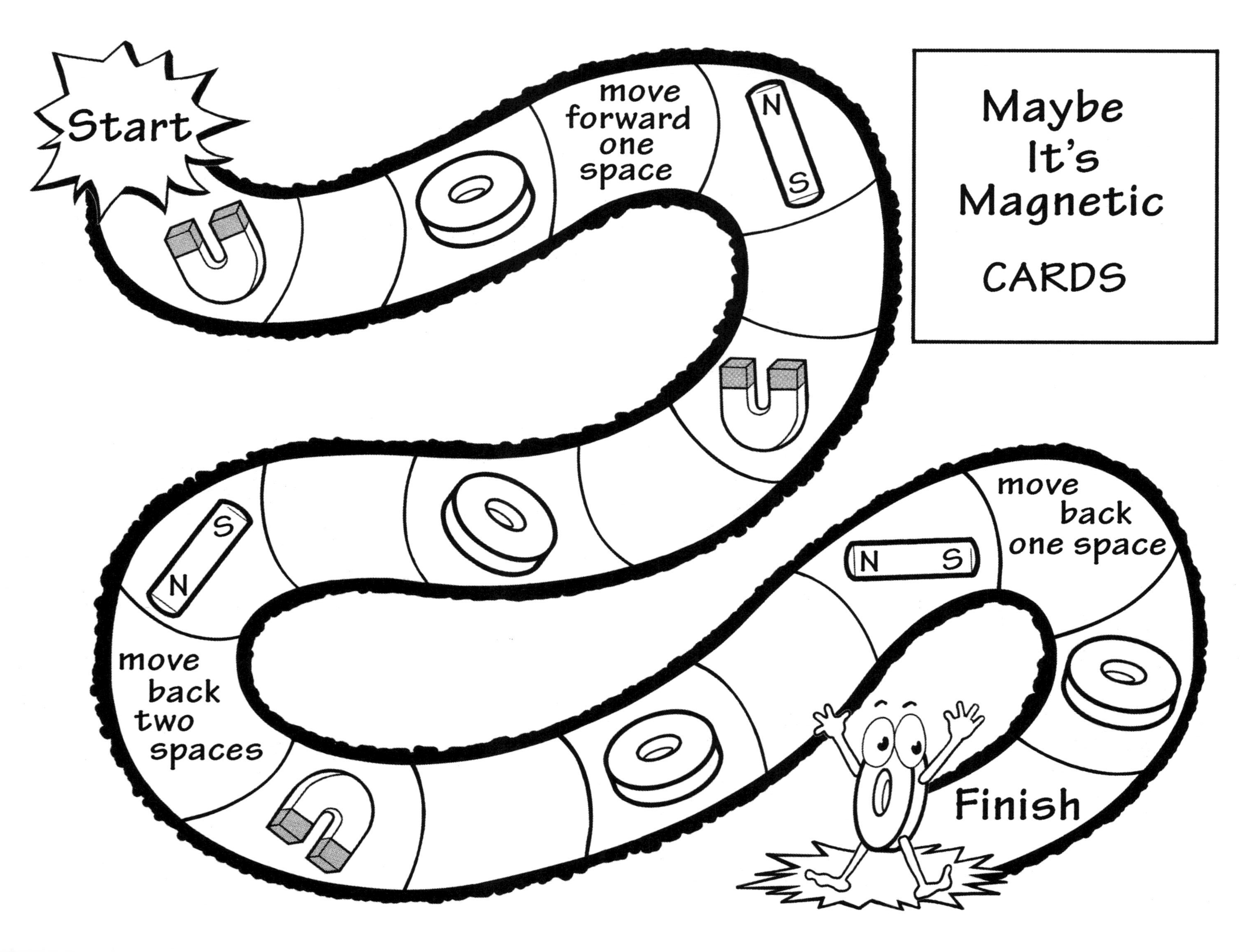
Start
move
forward
one
space
N
S
Maybe
It's
Magnetic
CARDS
S
N
move
back
two
spaces
N
S
move
back
one space
Finish

Topic
Magnetism

Key Question
Which magnet is the strongest?

Learning Goals
Students will:
- discover that magnets come in different sizes, shapes, and strengths; and
- test to see which magnet is "mightiest" (strongest).

Guiding Documents
Project 2061 Benchmarks
- *Without touching them, a magnet pulls on all things made of iron and either pushes or pulls on other magnets.*
- *Keep records of their investigations and observations.*

NRC Standards
- *Plan and conduct a simple investigation.*
- *Magnets attract and repel each other and certain kinds of other materials.*

*NCTM Standard 2000**
- *Represent data using concrete objects, pictures, and graphs*

Math
Graphing

Science
Physical science
force
magnetism

Integrated Processes
Observing
Comparing and contrasting
Classifying
Predicting
Recording

Materials
For each pair of students:
horseshoe magnet
bar magnet
ring magnet
paper clip
crayon
student page

Background Information
There are three main types of magnets—temporary magnets, electromagnets, and permanent magnets. Temporary magnets are those that act like a permanent magnet when they are within a strong magnetic field, but lose their magnetism when the magnetic field disappears. For example, paper clips and nails can become magnetically charged after repeated contact with a magnet, but the magnetic charge eventually wears off.

An electromagnet is a tightly wound coil of wire, usually with an iron core. It acts like a permanent magnet when electric current is flowing through the wire. When there is no current, there is no magnetic field.

The magnets with which we are most familiar are permanent magnets. They come in different sizes, shapes, and strengths and often appear on our refrigerators. They are considered permanent because once they are magnetized, they retain a level of magnetism. A magnet's strength is a measure of the magnet's force of attraction.

In this activity, students will test the strength of three types of permanent magnets—a bar magnet, ring magnet, and horseshoe magnet.

Management
1. If one of each type of magnet is not available for each group, have groups take turns with the available magnets, or place each in a center through which students can rotate.
2. The test page will create a graph of the results for students to interpret.
3. Ring magnets (order number 1971) and cow magnets (order number 1951)—a type of bar magnet—are available from AIMS.

Procedure

1. Discuss the concept that magnets come in different shapes and sizes. Ask students to identify the kinds of magnets they have seen and used.
2. Introduce the three different types of magnets that the students will be using in the activity.
3. Ask students to predict which magnet they think is the "mightiest" and tell why they made that prediction.
4. Explain that they will be testing the strength of each of the kinds of magnets by seeing how close the magnet will come to a paper clip before the paper clip is attracted to the magnet. The mightiest magnet will be the one that attracts the paper clip from the furthest distance.
5. Give each pair of students a paper clip, the student page, and one each of the three kinds of magnets. Explain that they will be testing each magnet separately. Tell them to place the paper clip at the top of the first column and to place the horseshoe magnet at the bottom of the first column so that it is just touching the line. Have them move the magnet closer to the paper clip one box at a time. Direct students to mark the column at the point where the magnet attracts the paper clip and color all boxes below that line.
6. Tell the students to tap the paper clip three times on the table top and to repeat the procedure using the other two magnets. (By tapping the paper clip, students will remove any remaining magnetic effect that is on it by being in contact with the magnet.)
7. Bring the class together to discuss and compare results. Encourage them to design other experiments that would show which is mightiest.

Connecting Learning

1. Which magnet did you predict would be the mightiest? Why?
2. Which magnet was actually the mightiest? How do you know? [It was furthest from the paper clip when the clip moved.]
3. Which magnet was the weakest? How do you know? [It was closest to the paper clip when the clip moved.]
4. What conclusions can we make from our tests?
5. How does the graph page help us interpret our results? [It shows us how the three magnets compare. You can see how much difference there is between the strongest and the weakest magnet. Etc.]
6. What other tests could we have used to decide which magnet was mightiest?
7. What are you wondering now?

Extension

Using one magnet at a time, place the magnet in a pile of washers. After removing magnet, count washers that remain attached to the magnet.

* Reprinted with permission from *Principles and Standards for School Mathematics*, 2000 by the National Council of Teachers of Mathematics.

Mightiest Magnet
S
N
Circle the mightiest magnet.

Topics
Magnetic force, attract and repel

Key Question
What happens when you put magnets together?

Learning Goals
Students will:
- determine the like poles on their magnets, and
- discover how like and unlike poles interact with one another.

Guiding Documents
Project 2061 Benchmark
- *Without touching them, a magnet pulls on all things made of iron and either pushes or pulls on other magnets.*

NRC Standard
- *Magnet attract and repel each other and certain kinds of other materials.*

Science
Physical science
force
magnetism
poles

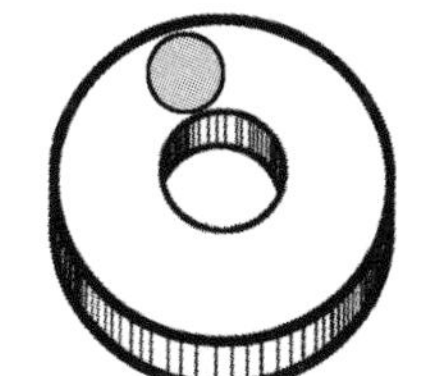

Integrated Processes
Observing
Predicting
Comparing and contrasting
Generalizing

Materials
For each student:
ring magnet
small adhesive dot
paper circle (see *Management 4)*
string, 50 cm (see *Management 4)*

Background Information
The areas of a magnet where magnetic force is most concentrated are its poles. Magnets usually have two poles. If two magnets are placed side by side with like magnetic poles together, they will repel one another. If their unlike poles are placed together, they will attract. Students learn that forces are pushes and pulls. Attraction and repulsion are the proper terms for these pushes and pulls when working with magnets.

Management
1. Each child will need a ring magnet. Ring magnets (order number 1971) are available from AIMS. Depending on the strength of the magnets you are using, you may need to warn children to be careful not to pinch their fingers when putting unlike poles together.
2. If small adhesive dots are not available, small pieces of masking tape can be used to mark like poles.
3. Allow students to have some free exploration time with the magnets. Caution them not to put the magnets near any electronic equipment such as computers or televisions.
4. Cut out large circles from paper that match the color of the adhesive dots you are using. Tape both ends of a string to each circle to form necklaces.

Procedure
1. After a period of free exploration time with the magnets, tell students that they are going to try to answer the question *What happens when you put magnets together?*
2. Inform them that they are going to mark their magnets so they can tell which side is being discussed.
3. Select a student to be the magnet stack starter (maybe a child whose birthday is closest). Tell students that they are going to put their magnets on top of the starter magnet.
4. As students file by to stack their magnets, use the terms *attract* and *repel* so learners will observe the interaction and begin to develop the proper vocabulary. Have the children stay in line. Make sure that all magnets are in the stack.
5. When the last magnet has been added to the stack, tell the children that they are going to pull their magnets off the stack, but before they do, they need to put a colored sticker on the top of the magnet.

6. As students reverse the order of their line, make sure that each child adheres a sticker dot to the top of the magnet before it is taken off the stack. This ensures that the like sides of all magnets are marked.
7. When all students have their marked magnets, invite them to go to other classmates and see what happens when they put their magnets together. They will need to be able to answer three questions: What happens when they put the sticker side of one magnet next to the sticker side of another magnet? What happens when they put the sticker side of one magnet next to the plain side of another magnet? What happens when they put the plain side of one magnet next to the plain side of another magnet?
8. Gather the class together to answer the three questions. Use the terminology *attract* and *repel.*
9. Invite two students to the front of the room. Have them put on the circle necklaces. Tell the class that they are going to pretend they are magnets. Since they are wearing the circles on the front side, this side will be like the magnet side with the sticker. Their backsides have no circles on them, so they will be like the magnet side without the sticker.
10. Call out "sticker to sticker" and have the two students move apart as though they are repelled. Call out "sticker to plain" and have the two students (standing front to back) move together as though they are attracted. Ask the class what the students should do when you call out "plain to plain." [Turn their backs to each other and move apart.]
11. When students understand what to do, distribute circle necklaces to each student. Have them pair up and practice attraction and repulsion. Fun can be added by having them move according to Simon Says rules. After each move, have students check with their magnets.

Connecting Learning

1. What happens when a magnet attracts another magnet? [They go together.]
2. What happens when a magnet repels another magnet? [They move apart.]
3. How were you able to get your magnets to attract each other? [We had to put a sticker side with a plain side.]
4. How many ways can you make magnets repel each other? [two] What are those ways? [sticker to sticker or plain side to plain side]
5. If someone told you that you could win a prize if you could stack one magnet on top of another without it being repelled, could you do it? How?
6. What are you wondering now?

Formative Assessment

1. Write the words *attract* and *repel* on separate sheets of papers.
2. Pair students and give each student a magnet with the stickers still on.
3. Review the words and their meanings. Tell the students that you are going to hold up one of the words and they must use their magnets to demonstrate it.
4. Watch pairs to see that students understand that like poles (sticker to sticker or plain side to plain side) repel while unlike poles (sticker to plain) attract.

Together Again

Topic
Magnetism

Key Question
Why do you think magnets are attracted to each other sometimes and at other times they don't want to go near each other?

Learning Goals
Students will:
- explore how like and unlike poles interact, and
- play a game to review that like poles repel and unlike poles attract.

Guiding Documents
Project 2061 Benchmark
- *Without touching them, a magnet pulls on all things made of iron and either pushes or pulls on other magnets.*

NRC Standard
- *Magnets attract and repel each other and certain kinds of other materials.*

Science
Physical science
  force
    magnetism
      attract, repel

Integrated Processes
Observing
Comparing and contrasting
Classifying
Predicting

Materials
For each student:
  bar magnet (see *Management 1)*
  construction paper
  40 cm of yarn
  marker
  sticky dots (see *Management 2)*

Background Information
Magnets have a magnetic field with the strongest magnetic forces at the poles. There are usually two poles on a magnet—north-seeking and south-seeking. Like poles repel each other and unlike poles attract each other.

In this activity, students will explore how magnets interact with each other. They will learn the terms that describe those interactions and will then apply what they know about the interaction of two magnets in a game.

Management
1. Bar magnets are suggested for this activity. Cow magnets are a type of bar magnet. They are available from AIMS (order number1951). If they are not available, ring magnets (order number 1971) will work, however, their poles are on the faces, not the ends.
2. Before beginning, mark the north (or south) pole on each magnet. To do this, line up the magnets so that they are attracted end to end. Place a small sticky dot or piece of masking tape on the same end of each magnet. Since opposites attract, the north-seeking (or south-seeking) pole will always be at the same end on each magnet.

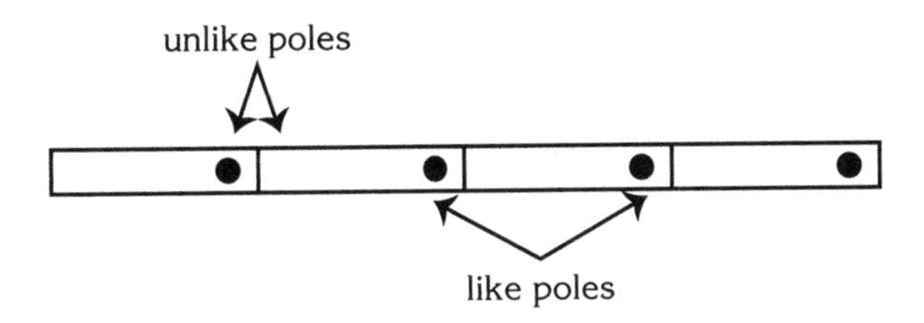

3. Students will work in pairs.

Procedure
Part One
1. Pair up students, giving each student a magnet. Ask them to explore how the magnets interact with each other. Encourage them to test all sides of the magnets.
2. After a time of free exploration, have the student pairs share their discoveries.

3. Ask the *Key Question.* Record student responses. Explain that there are science terms that describe how the magnets acted. Tell the students that one of the words is *attract.* Ask what they think it means for magnets to be *attracted* to something. Explain that the opposite of attract is *repel*, which means to push away. Show two magnets being attracted to one another. Demonstrate how you can make the same two magnets repel each other by turning one of them around.
4. Explain that every magnet has two poles and that the poles are on the ends of bar magnets.
5. Ask the pairs of students to try to put both poles marked with stickers together. Tell them that these poles are alike because they both have stickers. Discuss the results. [Like poles repel each other.]
6. Ask the groups to put one pole with a sticker and one without a sticker together. Tell them that these poles and not alike, or unlike, because one has a sticker and the other doesn't. Discuss the results. Make the generalization that unlike poles attract.

Part Two

1. Give each student a piece of construction paper, marker, and piece of yarn.
2. Instruct the students to draw a large rectangle on their papers. Tell them to write a large N or S on one end of the rectangle. Assist students in attaching the yarn to the paper to make necklaces.

3. Review the terms *attract* and *repel.* Explain to the students that they are doing to play a game where they are bar magnets and that when you say the word *attract* they are to quickly find a magnet that they would be attracted to and link arms. Remind students that unlike poles attract, so if they have written an "S" on their magnet, they will need to find someone with an "N." Invite a student to tell the class what they should do when you say *repel.* [Find a student with the same letter and stand and arm's length apart, pushing against each other's hands.] Tell the students that anyone without a partner is out and must sit down.
4. Take the class to a large open area and ask them to spread out. Remind them of the rules of the game. Try a practice round. Say the word *repel.* Students should find a partner with the same pole and they should stand at arm's length from them with their hands pushing their partner away.
5. Begin the game, calling our repel and attract. Play continues until time runs out or they are down to an even number of Ss and Ns.
6. End with a recap of how magnets interact with each other.

Connecting Learning

1. What is the word used to describe when magnets come together? [attract]
2. What is the word used to describe when magnets push apart? [repel]
3. If two like poles are end to end, will they repel or attract? [repel] How do you know?
4. What will happen when unlike poles are placed end to end? [They will attract each other.] How do you know?
5. When playing the game, what magnetic pole did you look for when I said *repel?* [same] ...*attract?* [different] Why?
6. What are you wondering now?

Topic
Magnetism

Key Question
How could we use a magnet to locate another magnet?

Learning Goal
Students will locate hidden magnets based on their reactions to other magnets.

Guiding Documents
Project 2061 Benchmarks
- *Without touching them, a magnet pulls on all things made of iron and either pushes or pulls on other magnets.*
- *Keep records of their investigations and observations.*

NRC Standards
- *Plan and conduct a simple investigation.*
- *Magnets attract and repel each other and certain kinds of other materials.*

Science
Physical science
force
magnetism

Integrated Processes
Observing
Comparing and contrasting
Predicting
Recording

Materials
Ring magnets
Student recording page
Milk cartons (see *Management 1)*
Permanent marker

Background Information

A magnetic force can't be seen; however, students can see evidence as the force of a magnet moves an object. In this activity, students will discover that magnetic forces can work from a distance without touching an object, and that they can work through materials, in this case a milk carton. Students will also apply their knowledge of how two magnets interact with each other through attraction or repulsion.

Management
1. Prior to teaching this lesson, prepare five milk cartons. These can be any size milk cartons from half-pint to half-gallon. Start by opening up the cartons so a magnet can be taped inside each one. Place the magnet in a different position in each carton. For example, in one carton, tape the magnet in the top right corner; in the next carton, tape a magnet in the middle of one side. When the magnets are in place, tape the cartons shut and use the permanent marker to label the cartons A through D and demo carton. Cut a piece of a sixth milk carton to use in demonstrating that a magnet will work through it. Number the sides of each milk carton one through four.

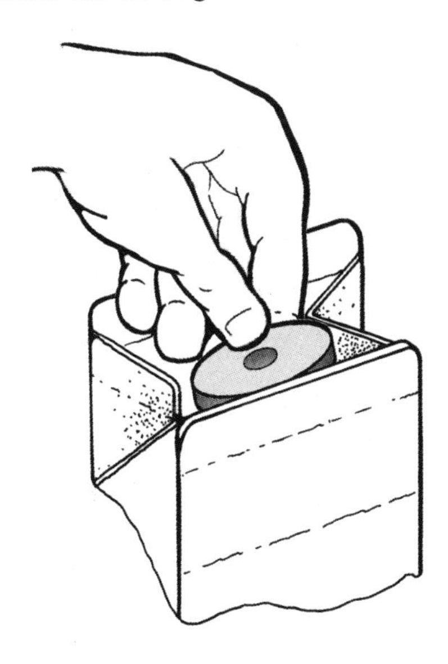

2. It is suggested that this activity be introduced to the entire class and then placed in a center for individual or small group work.
3. Ring magnets (order number 1971) are available from AIMS.

Procedure

1. Display the demo carton suggested in the *Management* section of this activity.
2. Explain to students that there is a magnet hidden in the carton and that you would like them to use their knowledge of magnets to help you find it.
3. Ask for suggestions as to how they might locate the hidden magnet. Accept reasonable suggestions and allow the student making the suggestion to test the idea.
4. If students do not have any suggestions, ask if they think a magnet could work through the milk carton. Invite a student to use a piece of milk carton and two magnets to find out.

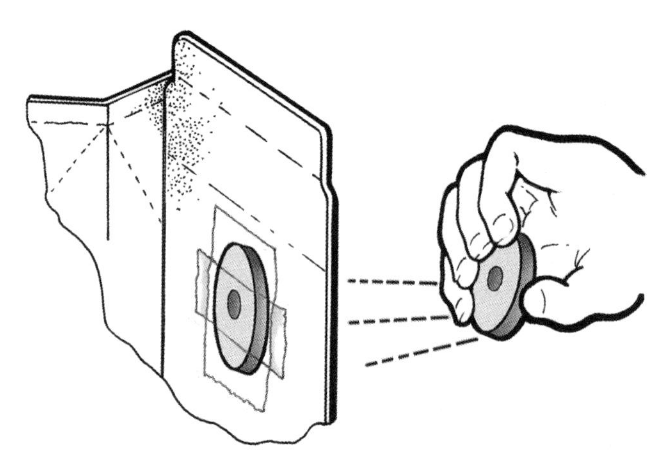

5. Show the students the other four milk cartons and the student recording page. Tell them that four milk cartons will be placed in a center and that throughout the week, they will rotate through the center. Remind them to record their results for later discussion.
6. When all students have had a chance to visit the center and find the hidden magnets, bring the class together and discuss their findings.

Connecting Learning

1. Did you find all of the magnets? How did you know when you located a hidden magnet?
2. Did the outside magnet always act the same when you waved it over the hidden magnet? Why?
3. Would a magnetic force work through water? How could you find out?
4. How could we find the hidden magnets if we didn't have a magnet to help us locate them? [We could use a paper clip; it would be attracted to the hidden magnets.]

Where are the magnets hidden?

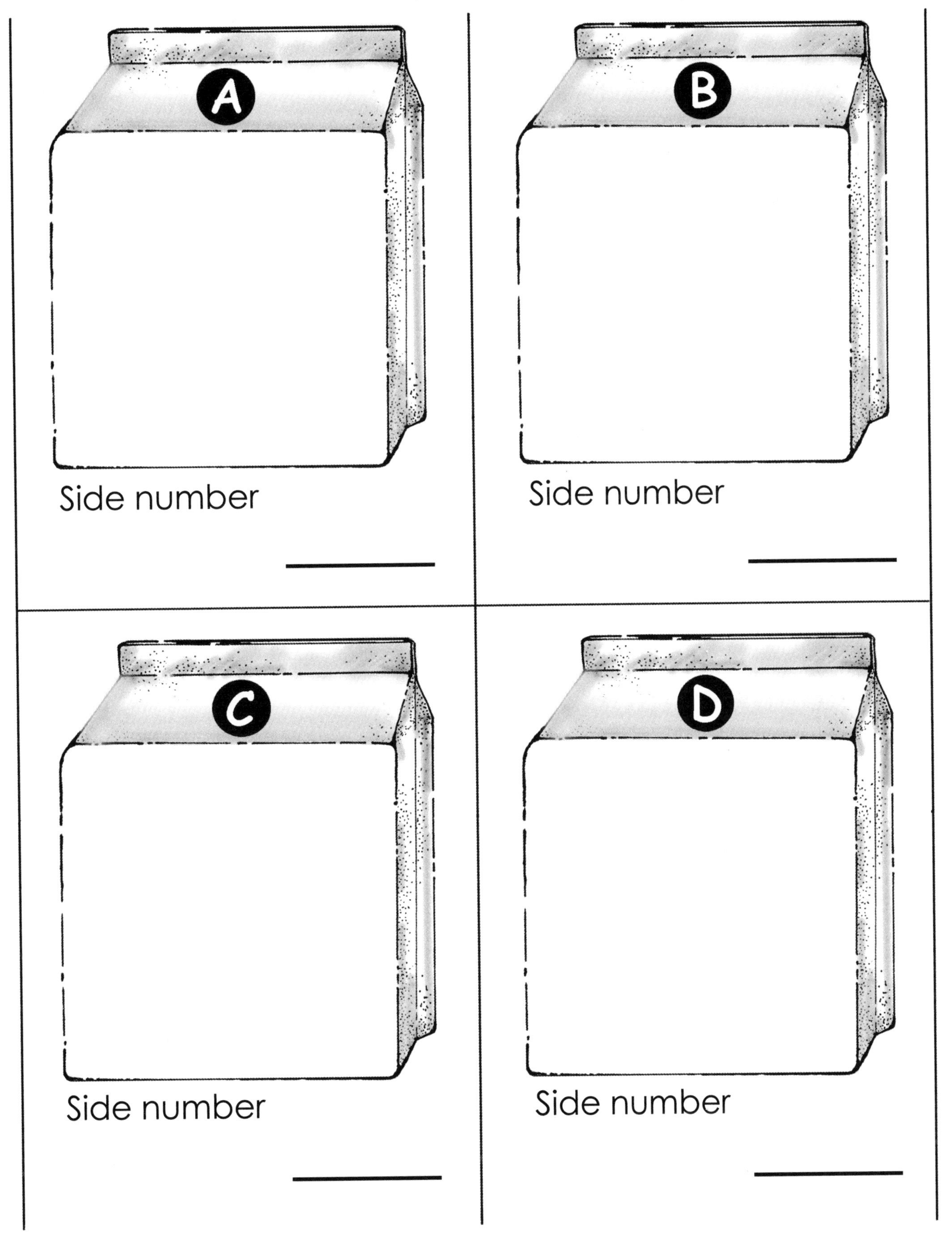

Topic
Magnetism

Key Question
How can you make an object move using a magnet?

Learning Goal
Students will move paper dogs along a path by pushing or pulling them with a magnet.

Guiding Documents
Project 2061 Benchmark
- *Without touching them, a magnet pulls on all things made of iron and either pushes or pulls on other magnets.*

NRC Standards
- *Magnets attract and repel each other and certain kinds of other materials.*
- *Employ simple equipment and tools to gather data and extend the senses.*

Science
Physical science
  force
    magnetism

Integrated Processes
Observing
Predicting
Interpreting data

Materials
For each student pair:
  two magnets
  ruler
  tape
  two paper clips
  Hungry Hound path page
  Hungry Hounds

Background Information
This investigation applies two basic characteristics of magnets, the attraction/repulsion of their poles and their ability to interact with magnetic objects through many materials. Every magnet has a magnetic field that is strongest at its poles. The poles of a ring magnet are on its large flat sides. If two magnets are placed facing each other with like magnetic poles together, they will repel one another; if unlike magnetic poles are together, they will attract. Magnetic fields are able to pass through many materials, demonstrated by placing a thin barrier, such as a piece of paper, between a magnet and a paper clip or between two magnets. Note that the magnet can only attract the clip, but two magnets can either attract or repulse one another. Attraction and repulsion would seem to be two types of magnetic force, but such is not the case. They are really two manifestations of a single magnetic force.

Management
1. Tape each magnet to the end of a ruler, as shown.

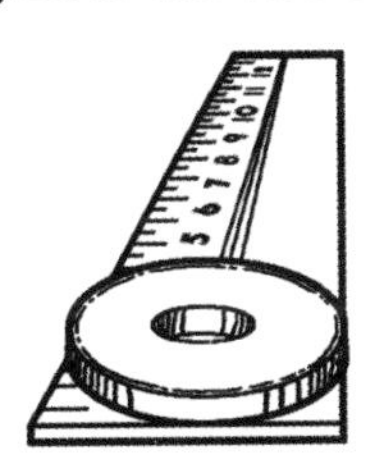

2. Assemble *Hungry Hounds* by taping tabs. Use separate tape for clips so the clips can be replaced with magnets for repelling.

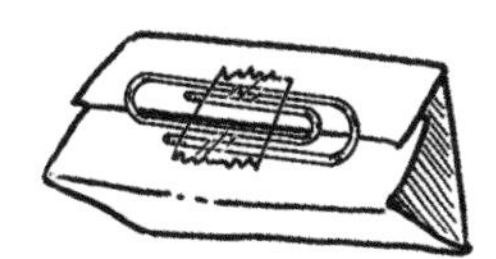

3. Keep it an exploratory activity rather than a competitive one.
4. Ring magnets (order number 1971) are available from AIMS.

Procedure
1. Distribute *Hungry Hounds* and activity sheets.
2. Ask the *Key Question*: How can you make an object move with a magnet? How could you make the *Hungry Hound* move down the path to the bone using a magnet?
3. Distribute rulers with magnets.
4. Have the students pull each hound along the path by placing the ruler under the track, so that the hound is on top of the magnet with the paper in between.

5. Ask students how they could *push* the hound. Replace the paper clip with a magnet. Check that the downward face of this magnet *repels* the upper face of the ruler magnet.
6. Have them push the hounds down the path (or try)!
7. Let students share what they observed.

Connecting Learning
1. What are two ways you can move an object? [by pulling it or by pushing it]
2. Was it easier to push or pull the hound down the path? Explain.
3. What else could we have used besides a paper clip?
4. What would happen if we turned over the hound's magnet?
5. What are you wondering now?

Extensions
1. Have students design other objects and tracks such as a rabbit and a path to some carrots or a boat with a water path.
2. Draw a larger set of paths on newsprint and use a meter stick instead of the ruler.
3. For more practice in visual-motor coordination, make an activity sheet with a dog's path that grows narrower as it approaches the bone, or try a race with two straight paths and two magnets, one taped on a ruler set to repel.
4. Have students try other barriers besides the paper (table, book, chair, etc.).

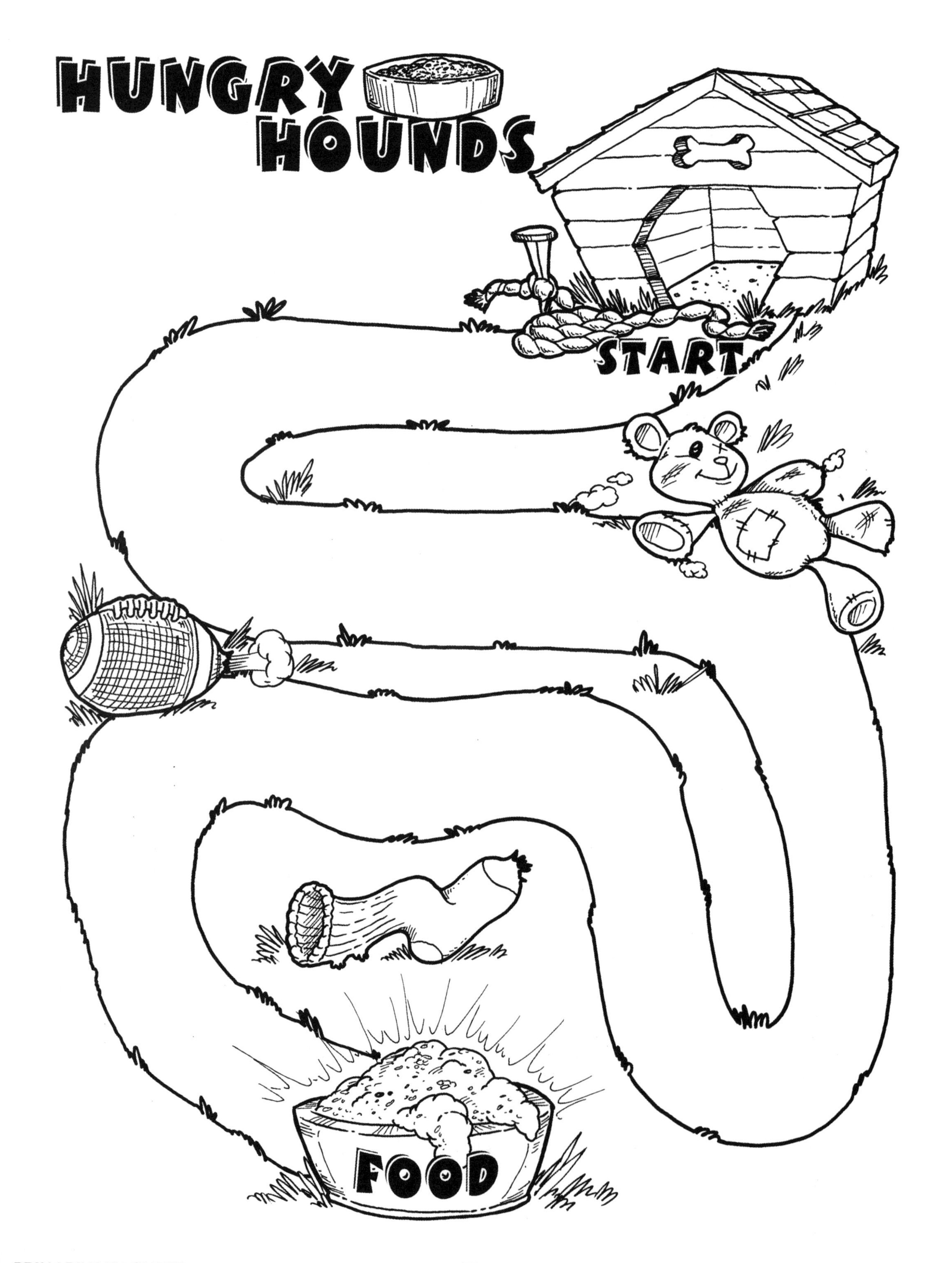
HUNGRY
HOUNDS
START
FOOD

Painting With Magnets

Topic
Magnetism

Key Question
How can you move a paper clip without touching it?

Learning Goal
Students will design a journal cover with a paper clip by pushing or pulling it through paint using a magnet.

Guiding Documents
Project 2061 Benchmark
- *Magnets can be used to make some things move without being touched.*

NRC Standard
- *Magnets attract and repel each other and certain kinds of other materials.*

Science
Physical science
 force
  magnetism

Integrated Processes
Observing
Applying

Materials
9- x 12-inch white construction paper, one per student
Ruler
Ring magnets
2 colors of tempera paint
Jumbo paper clips
Tape
Cardboard box lid (see *Management 1)*
White bond paper (see *Management 3)*
Stapler

Background Information
This investigation applies two basic characteristics of magnets—attraction of magnetic materials to a magnet and the ability to interact with magnetic objects through materials. The students can explore this concept by placing a piece of paper (non-magnetic) between a magnet and a small magnetic object (paper clip). Students should realize that the magnetic force goes through the paper and interacts with the steel paper clip that is attracted to the magnet. The students will use this attractive force to move the paper clip through paint to decorate a journal cover.

Management
1. The lid to a box of copy paper works well. If one is not available, choose any shallow box that will hold a 9" x 12" piece of paper so that the students can manipulate the magnet and see the paper and paper clip at the same time.
2. Divide the class into pairs. Each pair of students will need one ring magnet, one jumbo paper clip, a ruler, and a box lid.
3. Each student will need a sheet of white 9" x 12" construction paper and enough bond paper to make a journal.
4. Ring magnets (order number 1917) are available from AIMS.

Procedure
1. Give each pair of students a ring magnet and a jumbo paper clip.
2. Ask the *Key Question* and challenge students to try to move the paper clip without touching it. Give students time to experiment.
3. When each group has been successful at moving the paper clip using the magnet, tell the students that they are going to use what they have learned to paint a journal cover.
4. Distribute a ruler and tape to each group. Show students how to tape the magnet to the end of the ruler to make a magnetic "wand."

5. Give each group a box lid and each student one piece of construction paper. Tell one student in each pair to place the paper into the box along with the jumbo paper clip and practice moving the paper clip by putting the magnetic wand on the bottom of the box and moving it around.

6. Move from group to group with the paint and place a drop of each of two colors on the students' papers.
7. Invite the students to use the magnet to drag the paper clip through the paint to make colorful journal covers.

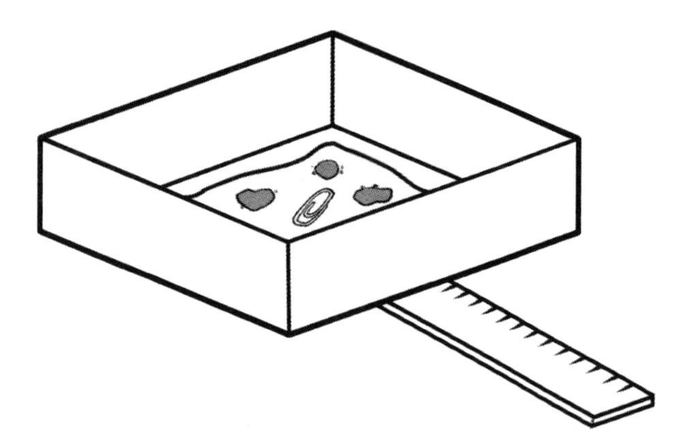

8. Allow time for the second student in each group to make a journal cover.
9. Once their paintings are dry, assist the students in constructing journals using their paintings as the cover and white bond paper for the inside pages.

Connecting Learning

1. Can a magnet move a paper clip? How do you know?
2. How can you move a paper clip without touching it?
3. What other things could you have used instead of a paper clip?
4. What are you wondering now?

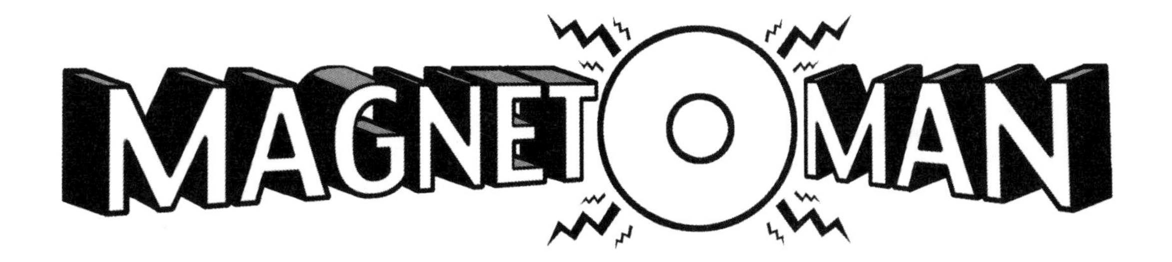

Topic
Magnetism

Key Question
How does magnet man work?

Learning Goal
Students will demonstrate the ability of magnets to attract and repel through certain materials.

Guiding Documents
Project 2061 Benchmark
- *Magnets can be used to make some things move without being touched.*

NRC Standard
- *Magnets attract and repel each other and certain kinds of other materials.*

Science
Physical science
  force
    magnetism

Integrated Processes
Observing
Drawing conclusions
Generalizing

Materials
Pint-size recloseable plastic bags, one per student
Iron filings (see *Management 1)*
Magnets
Transparent tape
Commercially-made magnet face board (see *Management 2)*
Picture of magnet man (see *Management 4)*

Background Information
This investigation allows students to explore two basic characteristics of magnets, attraction/repulsion and the ability to interact with magnetic objects through materials such as plastics. The students will describe how the commercially-made magnet face boards work and will create their own versions of the game.

Management
1. Iron filings are available at most teacher stores. Prepare a reclosable plastic bag containing 1 tablespoon of iron filings for each student. Seal the plastic bags with clear tape for safety purposes.
2. The commercially-made magnet face board starts with a blank face and metal shavings are moved around with the magnetic stick to create different looks. These can be purchased at most toy stores.
3. Warning—Keep all magnets away from computers.
4. Copy one face picture per student onto card stock and laminate for extended use.
5. Ring magnets (order number 1971) are available from AIMS.

Procedure
1. Show the class a commercially-made magnetic face board. Demonstrate how to use a magnet to move the filings to give the illusion of hair on the character inside. Ask the class to describe how they think the toy works. Tell the class that they are going to make their own toys similar to the one displayed.
2. Give each student the reclosable plastic bags containing 1 tablespoon of iron filings, a face picture, and a magnet.
3. Instruct the students to tape the picture of magnet man onto the plastic bag so that students can see the picture when looking through the bag.
4. Encourage the students to give magnet man hair using the magnet. Discuss why it works and how the magnet can move the filings through the plastic.

Connecting Learning
1. What can a magnet do?
2. What would happen if you put a plastic toy near a magnet?
3. What would happen if you put metal near a magnet?
4. Through what things can a magnet attract/repel?
5. Explain how the magnet man works.
6. What other games could you make using a magnet?

Extension
Explore other thicknesses and materials that a magnet can attract/repel through, such as a table.

Topics
Magnetism, Motion

Key Question
What are the ways that Willy the worm moves as he goes through the tunnels to his home?

Learning Goals
Students will:
- illustrate Willy's worm home by listening to a description of that home,
- compare their illustrations to the picture of the actual home, and
- use a magnet to move Willy through the tunnels to his home.

Guiding Documents
Project 2061 Benchmarks
- *Things move in many different ways, such as straight, zigzag, round and round, back and forth, and fast and slow.*
- *Without touching them, a magnet pulls on all things made of iron and either pushes or pulls on other magnets.*

NRC Standard
- *Magnets attract and repel each other and certain kinds of other materials.*
- *An object's motion can be described by tracing and measuring its position over time.*

Science
Physical science
  motion
  force
    magnetism

Integrated Processes
Observing
Identifying
Comparing and contrasting
Recording

Materials
For each group:
  magnet
  ruler
  tape
  chenille stem (see *Management 2)*
  student pages (see *Management 3)*

Background Information
This activity has dual foci, magnetism and motion. The metal in the center of the chenille stem is attracted to a magnet. This attraction is strong enough that the chenille stem can be moved by a magnet through a piece of paper.

Students will experiment with a variety of motions—straight, zigzag, back and forth, and round and round. They will be challenged to interpret the meanings of these words by drawing what they think the home of Willy the worm looks like. By comparing their drawings to those of their classmates and to the picture of Willy's actual home, the students will be able to refine their understandings of these terms.

Management
1. Tape each magnet to the end of a ruler, as shown. Ring magnets (order number 1971) are available from AIMS.

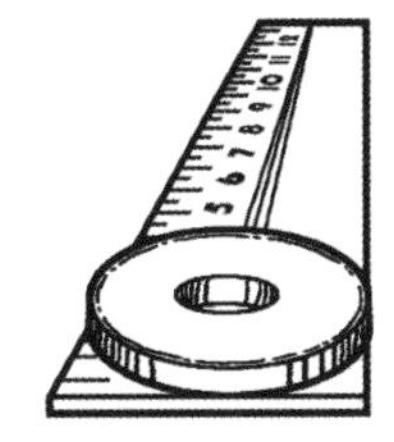

2. Cut chenille stems into fourths so that each student gets one three-inch section.
3. The page with the round-and-round section must be cut in half and taped to the other page to make one large picture. Assemble these pages before distributing them to students. These pages will be taped to the desk so that students can use the magnet under the paper to move the chenille stem "worm" through the tunnels. (See *Procedure 5.)*

Procedure

1. Tell students that you have a story for them about Willy the worm and that you would like them to listen carefully to the details about his home. Read the following:

   Willy the worm lives underground. He has a very fancy worm hole. At first, it was simple, but then he kept adding on. First, he tunneled straight down and made his home. That was good for a while, but Willy got bored with a straight tunnel. Every day it was back and forth in a straight line to get to the surface of the dirt. One day he decided to add on. He tunneled deeper into the soil. He zigzagged back and forth and back and forth and made a new home at the end of the zigzag tunnel. To get to the surface, he went back and forth and back and forth and then straight up through the original tunnel. A little later, he was getting restless. He decided to make his home even bigger. He tunneled round and round and round, down deeper into the ground. At last he had a home that he liked. Now he goes round and round, and back and forth, and straight to get to the surface.

2. Distribute the first student page and tell the students that you will read the story again slowly, and as you do so, they must draw how they think Willy's home looks based on the descriptions.
3. Read the story again and give students time to draw Willy's home as they think it looks. Invite them to share their drawings and observe the similarities and differences among the drawings.
4. Tell students that they will now get a page showing what Willy's home actually looks like. They will get a "worm" to move through the tunnels from the surface to the home.
5. Distribute the worm home page to each group of students and assist them in taping one edge of the worm home page to the desk. For right-handed students, tape the left side of the paper. For left-handed students, tape the right side of the paper. Invite students to compare Willy's actual worm home to the ones they drew. What are the similarities and differences?

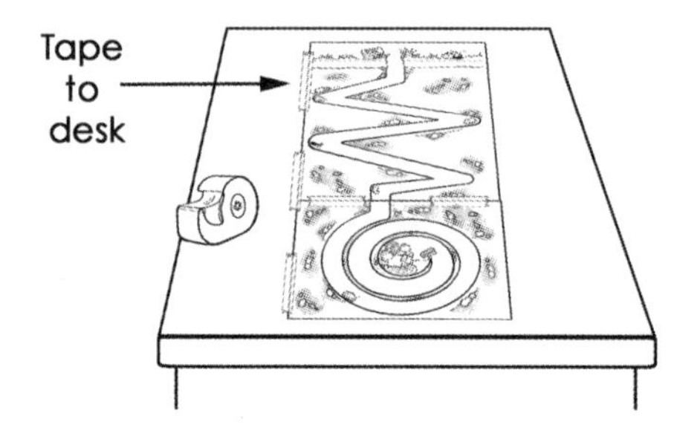

6. Give each group a chenille stem piece. Invite them to bend and/or twist the chenille stem to make a "worm."
7. Have the students take turns pull their worms through the tunnel by placing the ruler under the paper so that the worm is above the magnet with the paper in between.
8. Discuss again how Willy the worm moves as he goes through his home.

Connecting Learning

1. How did you decide what to draw when you were illustrating Willy's worm home?
2. What does it mean for something to zigzag or go back and forth?
3. What does it mean for something to go round and round?
4. Did everyone's drawings look the same? Why or why not?
5. How similar was the drawing you made to the way Willy's home actually looks?
6. What are some of the ways that Willy the worm moves?
7. What are you wondering now?

Extensions

1. Have students pretend to be Willy the worm and move through the classroom the way that Willy moves through his home.
2. Allow students to design their own worm homes and describe the movements needed to go through the tunnels.

Willy the Worm
WILLY'S PLACE

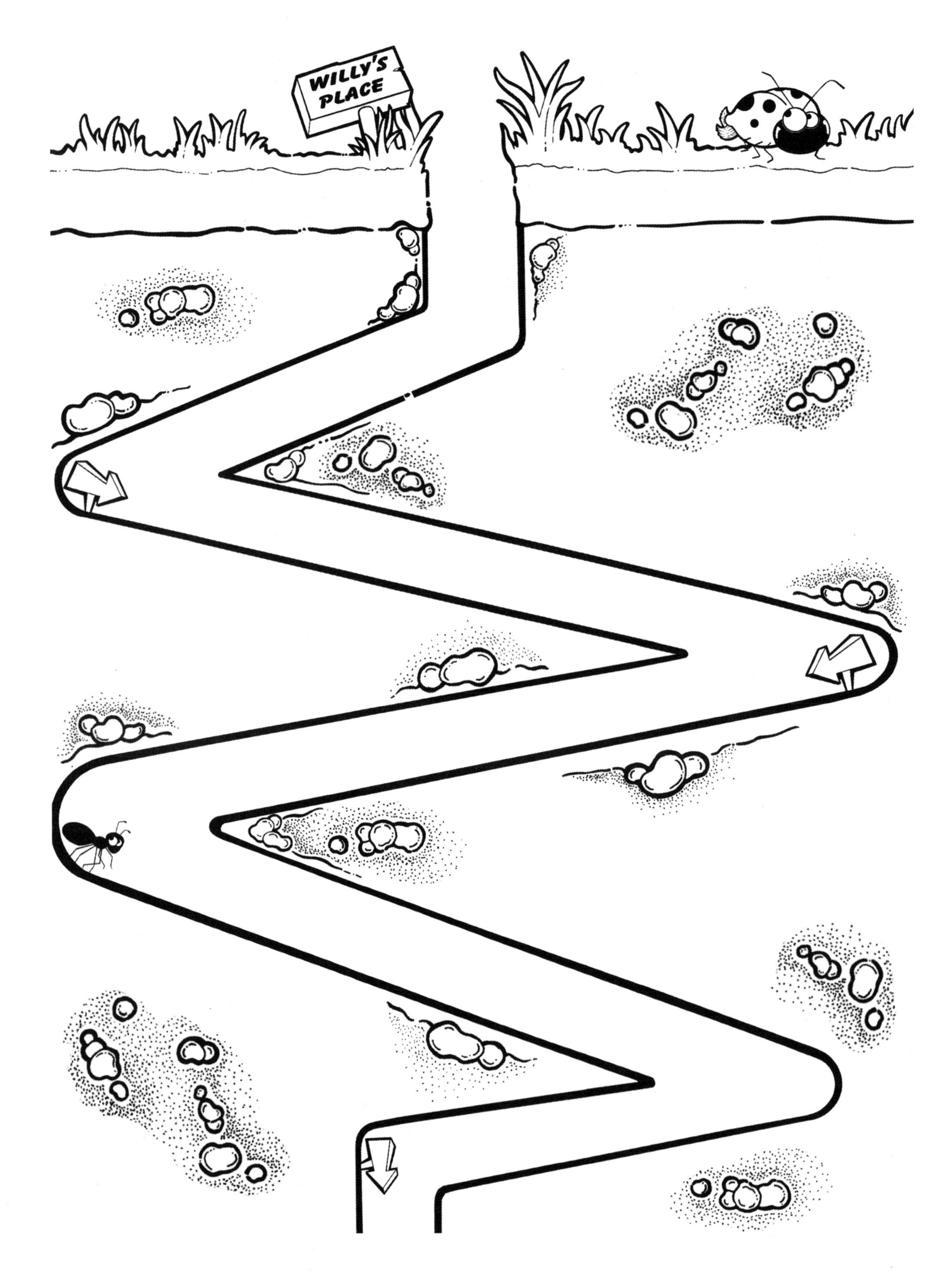
WILLY'S
PLACE

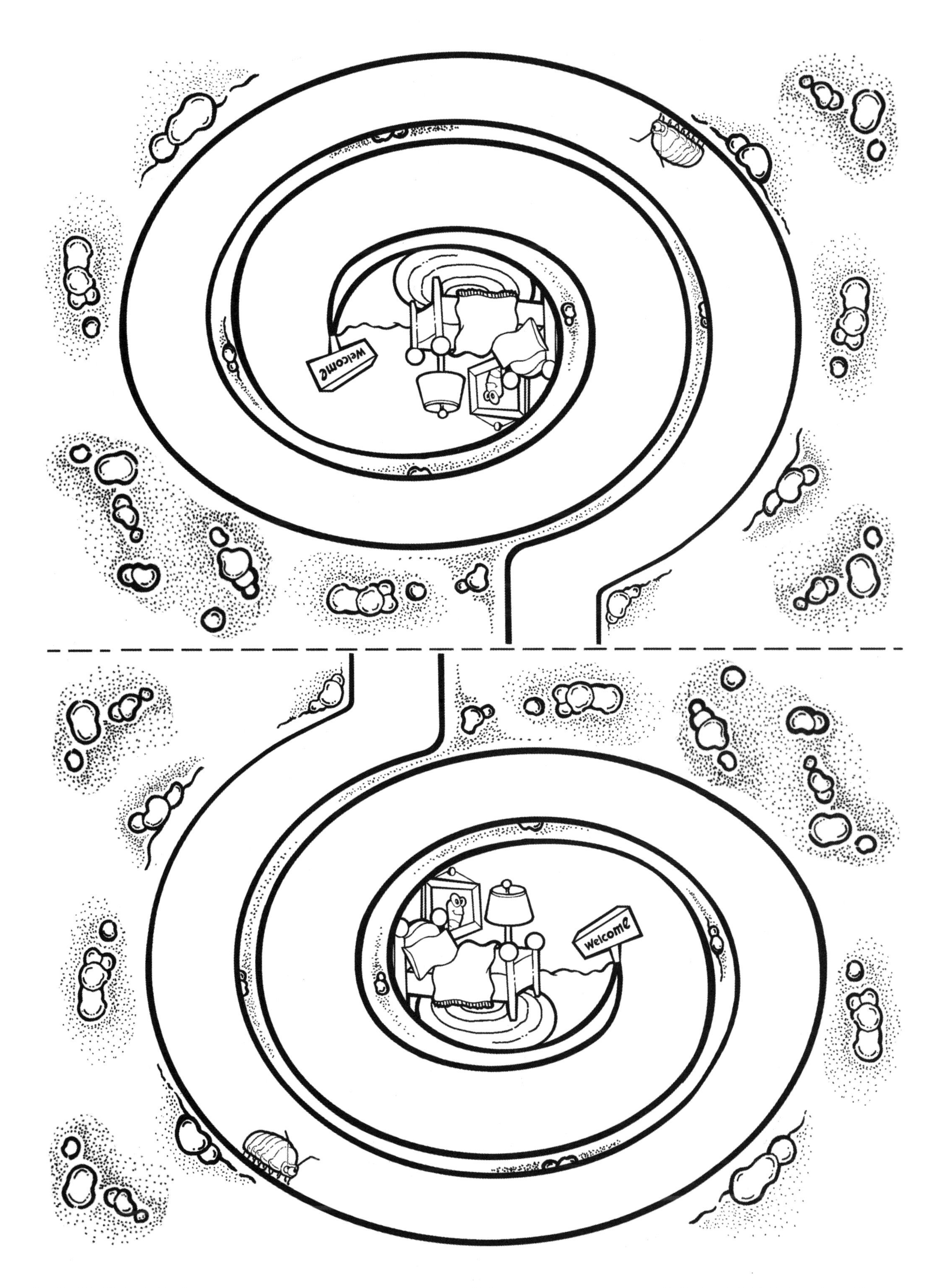
Welcome
Welcome

Make It Fly

Topic
Magnetism

Key Question
How can we use a magnet to fly a model kite?

Learning Goal
Students will "fly" a model kite by applying their knowledge of magnets and how they can work without touching an object.

Guiding Documents
Project 2061 Benchmarks
- *Without touching them, a magnet pulls on all things made of iron and either pushes or pulls on other magnets.*
- *Keep records of their investigations and observations.*

NRC Standards
- *Plan and conduct a simple investigation.*
- *Magnets attract and repel each other and certain kinds of other materials.*
- *The position and motion of objects can be changed by pushing or pulling. The size of the change is related to the strength of the push or pull.*

Science
Physical science
  force
    magnetism
    gravity

Integrated Processes
Observing
Comparing and contrasting
Predicting
Recording
Generalizing
Applying

Materials
Magnets (see *Management 1)*
Yarn (see *Management 2)*
Large paper clips
Tape
Sticky notes, 3 in x 3 in (see *Management 3)*

Background Information
The forces of gravity and magnetism are explored in this activity. Gravity is a force of attraction between objects without making direct contact. Its attraction is what keeps us from flying off the surface of the Earth. Magnetism is a force that we can't see but can experience. Like gravity, the magnetic force can act at a distance as a non-contact force. In this activity we will use the attraction of a magnet to a paper clip to overcome gravity and "fly" a model kite.

Management
1. Ring magnets (order number 1971) or cow magnets (order number 1951) can be used for this activity. Both are available from AIMS.
2. Cut yarn into 50-centimeter lengths. Each student needs one length.
3. Each student needs two sticky notes.

Procedure
1. Ask the *Key Question.*
2. Build a kite while the students watch. With the non-sticky side of the notes up, draw two diagonal lines on both the sticky notes. Turn one sticky note over so the sticky side is up. Tape a large paper clip in the middle of this sticky note. Tape a piece of yarn to one corner of the sticky note. Place the second sticky note on top, with the sticky end on the opposite end of the first sticky note. The two notes should stick together on two ends. Tape the loose end of the yarn to the table.

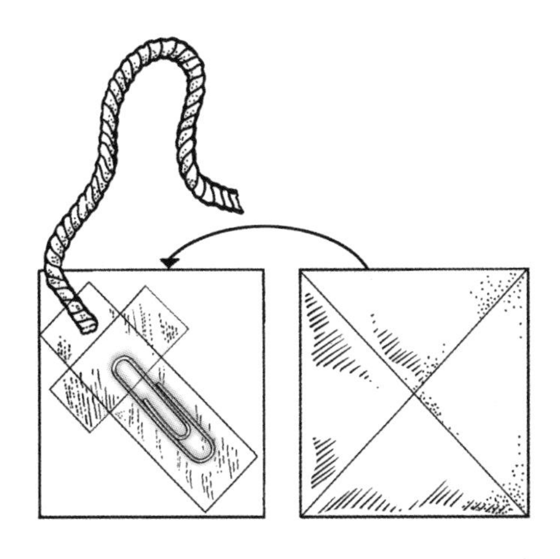

3. Hold your kite for students to see. Ask them to predict what will happen if you let go of it. Discuss the force of gravity acting on the kite unless something holds it up.
4. Remind students to keep magnets away from electronic equipment such as computers, DVD players, watches, and VHS tapes.
5. Give students the materials to make a kite and a magnet. Challenge them to get their kite to fly without touching it.
6. Provide time for students to work on their challenge. When all students have been successful, discuss what they did to make their kite fly. Ask them to include how they overcame gravity and why their solution worked.

Connecting Learning

1. What is gravity? How did we overcome gravity to fly our kites?
2. How did your kites work? What did you have to know about magnets to get it to work? What did you have to know about gravity?
3. What could you use instead of a paper clip? Would a piece of plastic work? Why?
4. What other toys could you make with a magnet?

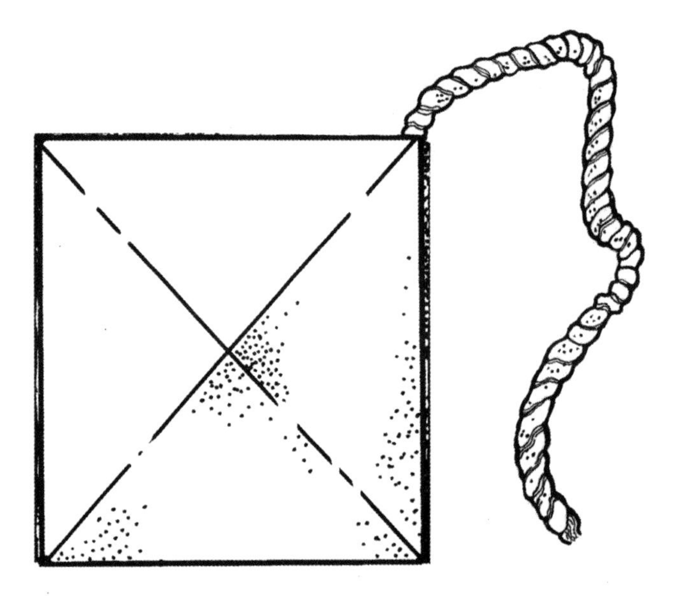

Topic
Magnetism

Key Question
What kinds of magnets are in our homes and school?

Learning Goal
Students will:
- identify everyday uses of magnets, and
- make their own refrigerator magnets.

Guiding Documents
Project 2061 Benchmark
- *Magnets can be used to make some things move without being touched.*

NRC Standard
- *Magnets attract and repel each other and certain kinds of other materials.*

Science
Physical science
force
magnetism

Integrated Processes
Observing
Identifying

Materials
Cookie sheets
Acrylic paint—yellow, white, black, peach
Paintbrushes
Acrylic spray sealer
Magnets
White craft glue
Flour
Salt
Water
Measuring cups
Aluminum foil
Toothpicks
Ring magnets, one per student

Background Information

Magnets have a variety of uses from home to medical to industrial. For example, magnetic resonance imaging (MRI) machines use strong magnetic fields to help doctors examine patients' internal organs. Videotapes and audiotapes store sound and image information on magnetic tape. Credit cards and ATM cards also use a magnetic strip that contains the information needed to access accounts. Microphones and speakers use a combination of permanent magnets and electromagnets to make them work. Older television sets and computer monitors use magnets to generate images. Powerful electromagnets are used in scrap yards to move large, heavy magnetic objects. They are also used to load and unload ships, barges, rail cars, and trucks.

Magnets are used to protect the health of animals. Cows are susceptible to a condition called traumatic reticulopericarditis, or hardware disease, that comes from swallowing metal objects (baling wire, barbed wire, staples, nails, etc.). These objects can puncture a cow's stomach and damage its diaphragm or heart, so veterinarians or ranchers use a special tool called a pill gun to insert magnets into back of a cow's throat. The cow swallows the magnet which stays in its first stomach. These long, narrow magnets, known as cow magnets, attract pieces of metal and prevent them from continuing on through a cow's system.

In the home, uses for magnets include electric can openers, refrigerator magnets, and magnetic screwdrivers.

Management
1. The salt-flour mixture can be made in class or ahead of time and stored in a re-closable plastic bag.
2. Salt-flour dough is one part salt, one part water, and two parts flour. Dough should not be sticky; you should be able to work it in your hands. If it is too sticky, add small amounts of flour until you get the consistency that you want. Be careful not to add too much flour, as the dried form will crack.
3. The salt-flour magnets can air dry, but it may take several days. It is suggested that you make arrangements with the cafeteria staff to bake them or take them home to bake.

4. The directions are given for students to make smiley magnets; however, you may choose to allow them to make a design of their own. Additional items such as beads can be added.
5. Ring magnets (order number 1971) are available from AIMS.

Procedure

1. Discuss ways that magnets are used in industry, electronics, and the medical field, and even in our homes. (See *Background Information.)*
2. Give students some examples of magnets in your own home, such as magnets on cabinet doors, refrigerator magnets, a magnetic screwdriver, can opener, etc.
3. Ask students, "Do we have any magnets in our classroom?" Give students time to look around the room before answering. Record student responses. Some examples of magnets may be magnetic letters or numbers, a magnetic paper clip holder, a magnetic white board, etc.
4. Tell the class that one common use for magnets is to hold things onto the refrigerator. Explain that they will be making a refrigerator magnet out of a salt-flour dough mixture.
5. Make the dough as a class or bring out the prepared dough.
6. Give each student a walnut-sized piece of dough and a four-inch square of foil on which to work. Ask them to form a ball by rolling the dough in their hands. Have them flatten the ball of dough onto the foil; it should be about ½-inch thick.
7. Give each student a small piece of dough with which to form a round nose and two small round eyes.
8. Have them use a toothpick to draw on the smiles. (Be sure to do this step before baking). Encourage students to personalize their smiley faces by adding hair, glasses, or freckles.
9. Place the students' dough faces on cookie sheets and bake them for 2½ hours at 250 degrees.
10. Remove from the oven and let cool completely. When the smiley faces have cooled, instruct students to paint the faces yellow and the noses peach. Let the paint dry, then repeat for a second coat. Have them use the handle end of the paintbrush to add the whites of the eyes. Let it dry completely before using a toothpick to add the blacks of eyes. Tell students to use the toothpick dipped in black to paint the smile lines.
11. Spray each smiley face with acrylic sealer spray.
12. Assist students with gluing a magnet to the back and let dry.
13. End with a review of how magnets are used in everyday life.

Connecting Learning

1. Name some ways that magnets are used.
2. How do we use magnets at school?
3. Why do you think magnets are important?
4. What magnets do you have at home?
5. What are you wondering now?

Topic
Magnetism

Key Question
How do magnets help us with everyday tasks?

Learning Goal
Students will identify everyday uses of magnets.

Guiding Documents
Project 2061 Benchmark
- *Magnets can be used to make some things move without being touched.*

NRC Standard
- *Magnets attract and repel each other and certain kinds of other materials.*

Science
Physical science
force
magnetism

Integrated Processes
Observing
Comparing and contrasting
Predicting

Materials
Crayons
Pictures and/or items containing magnets
Scene containing magnetic items, included

Background Information
Magnetism is a non-contact force. Magnets can cause some objects to move toward them. Magnets are used in many items that we use every day. When dealing with magnets, you need to observe certain safety precautions:
- Do not place magnets near mechanical watches, hearing aids, pacemakers, credit cards, TVs, audio or video tapes, or computer screens.
- Don't drop them. They may break and the impact will affect the strength of the magnet.

Management
1. The students may be able to come up with several uses of magnets. In case they can't, some to keep in mind are: magnetic door catches; magnetic purse catches; magnetic strips on credit cards; refrigerator magnets; magnets in toys, can openers; compasses; magnetic screwdriver; etc.

Procedure
1. Ask the class what they can tell you about magnets. Record their responses on chart paper under the heading, *What we know about magnets.*
2. Ask students to brainstorm ways we use magnets in our everyday lives. Allow a few minutes of think time before accepting responses. Record correct responses on chart paper under the heading *Everyday uses for magnets.*
3. Show students items or pictures of items that we use that contain magnets. As you display each item, ask students to predict where the magnet is.
4. Distribute the picture page and have students share the examples of magnets they find. When correct items are identified, have students circle them on their paper.
5. End with a discussion about how magnets help us and what invention we wish someone would make that would use magnets in some way. For example, magnetic shoe clasps so that we wouldn't have to tie our shoes.

Connecting Learning
1. What do you know about magnets?
2. Name some everyday objects that have magnets in them.
3. How can magnets make things move without being touched?
4. How is the magnet in a can opener helpful?
5. What are you wondering now?

Extension
Direct students to look at home for at least two ways that their family uses magnets at home or work. Have students write and/or draw to record the uses.

A Partial Key for the Picture
refrigerator magnet, can opener, magnetized knife holder, latch on microwave, purse clasp, credit card

Signature
Bank of USA
STAR
BINGO
FREE
Cat

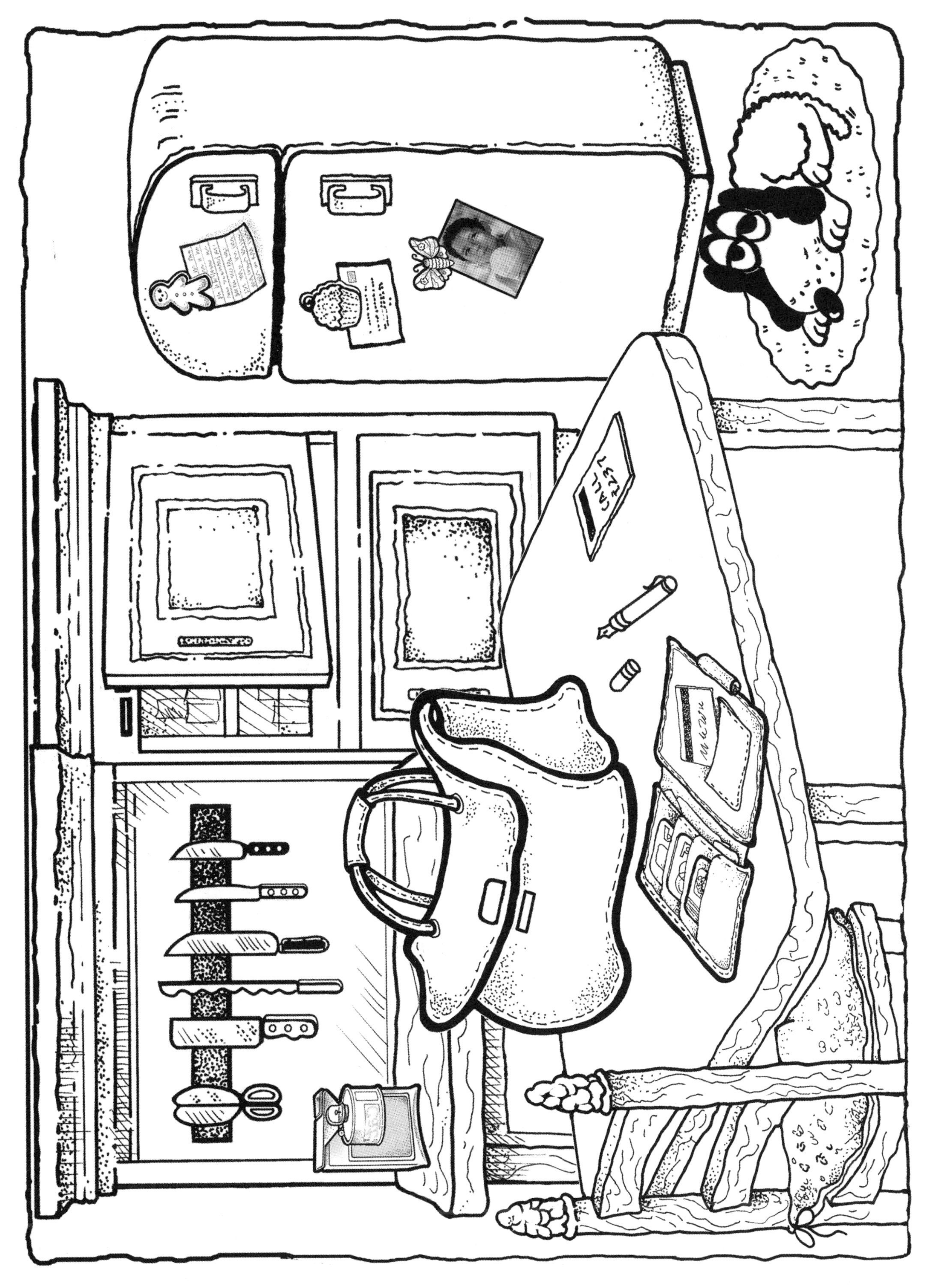

Topic
Magnetism

Key Question
How are electromagnets used in the real world?

Learning Goals
Students will:
- learn about electromagnets, and
- simulate the electromagnet used to move cars in a junkyard.

Guiding Documents
Project 2061 Benchmark
- *Magnets can be used to make some things move without being touched.*

NRC Standard
- *Magnets attract and repel each other and certain kinds of other materials.*

Science
Physical science
force
magnetism

Integrated Processes
Observing
Comparing and contrasting
Predicting

Materials
For each group of students:
3 large paper clips
3 small paper clips
30 cm of string
tape
ruler
3 ring magnets
1 card stock target
three card stock cars

Background Information
One of the key discoveries of all time was that electricity and magnetism are closely connected. Moving electric charges (an electric current) create a magnetic field. Conversely, a moving magnetic field creates an electric current in conducting materials. The discovery of this electromagnetic connection in the 19th century led to the invention of the electromagnet, the electric motor, and the electric generator.

Magnetic materials like iron that are not permanent magnets become temporary magnets in the presence of a magnetic field. As long as the magnetic field is present, the material acts like a magnet. When the magnetic field is removed, the material no longer acts like a magnet.

Since an electric current produces a magnetic field, coiling wire around a piece of iron can make an electromagnet. When current flows through this coil, it produces a magnetic field that turns the iron into a magnet. As soon as the current stops, the magnetic field is no longer present.

Electromagnets play an important role in technology. Large, powerful electromagnets attached to cranes are used in factories and junkyards to lift and move heavy pieces of steel or iron. Smaller electromagnets are utilized in the electromagnetic switches called solenoids. These switches are used in cars, doorbells, security doors, and many other mechanical devices.

In this activity, students will discuss various ways that magnets are used in our everyday lives and will simulate how electromagnets are used to move heavy objects such as cars. It is important to explain to students that the action is only being simulated. In real life, an electromagnet would be used to move heavy items and the flow of current through the electromagnet would be stopped to release the vehicle. In this simulation, a permanent magnet is being used and the magnetic attraction cannot be stopped to release the car. Instead, a greater magnetic force is created on the ground so that the car will be pulled off the end of the "crane."

Management
1. Copy the car and target pages onto card stock. To integrate math skills, cars and targets can be given number values and students can earn points for managing to get certain cars to certain targets.
2. To assemble the "crane," tie one end of the string to the magnet and tape the other end to the end of a ruler.

3. To set up the target, tape a two-unit magnet to the table, roll a piece of tape, and attach a card stock copy of the bull's-eye to the top of the magnet unit.
4. If the car is not releasing well, place the small paper clip on the under side of the card stock. It will cause the attraction to be weaker.
5. It is suggested that students work in small groups. However, the activity can be done as a whole class lesson or placed in a center as well.
6. Ring magnets (order number 1971) are available from AIMS.

Procedure

1. Ask the class if they have ever seen a cartoon where a big magnet lifts and moves something like a car or other large object. Question them about whether they think a large magnet like that exists and whether they that it could really lift a car.
2. Introduce the term *electromagnet.* Provide a simplified explanation of how an electromagnet works and how it is used in junkyards and shipyards to move heavy metal things. (See *Background Information.*)
3. Explain to the class that they are going to get to act like junkyard workers and move several "cars."
4. Divide the class into small groups. Give each group the materials needed.
5. Assist students in the construction of the vehicles and crane. Demonstrate how they are to pick up and deliver the cars to the target. Allow time for students to work with their magnets and "cars."
6. End with a discussion about how this experience simulated a real-world use of electromagnets.

Connecting Learning

1. How are electromagnets used every day?
2. How was our simulation similar to a real crane moving cars? How was it different? [A real crane would use an electromagnet, not a permanent magnet.]
3. What else could we have used instead of card stock cars with paper clips attached?
4. Do you think that real toy vehicles would be attracted to a magnet? How could we find out?
5. What are you wondering now?

Extension

Have students use the Internet to find other uses for electromagnets.

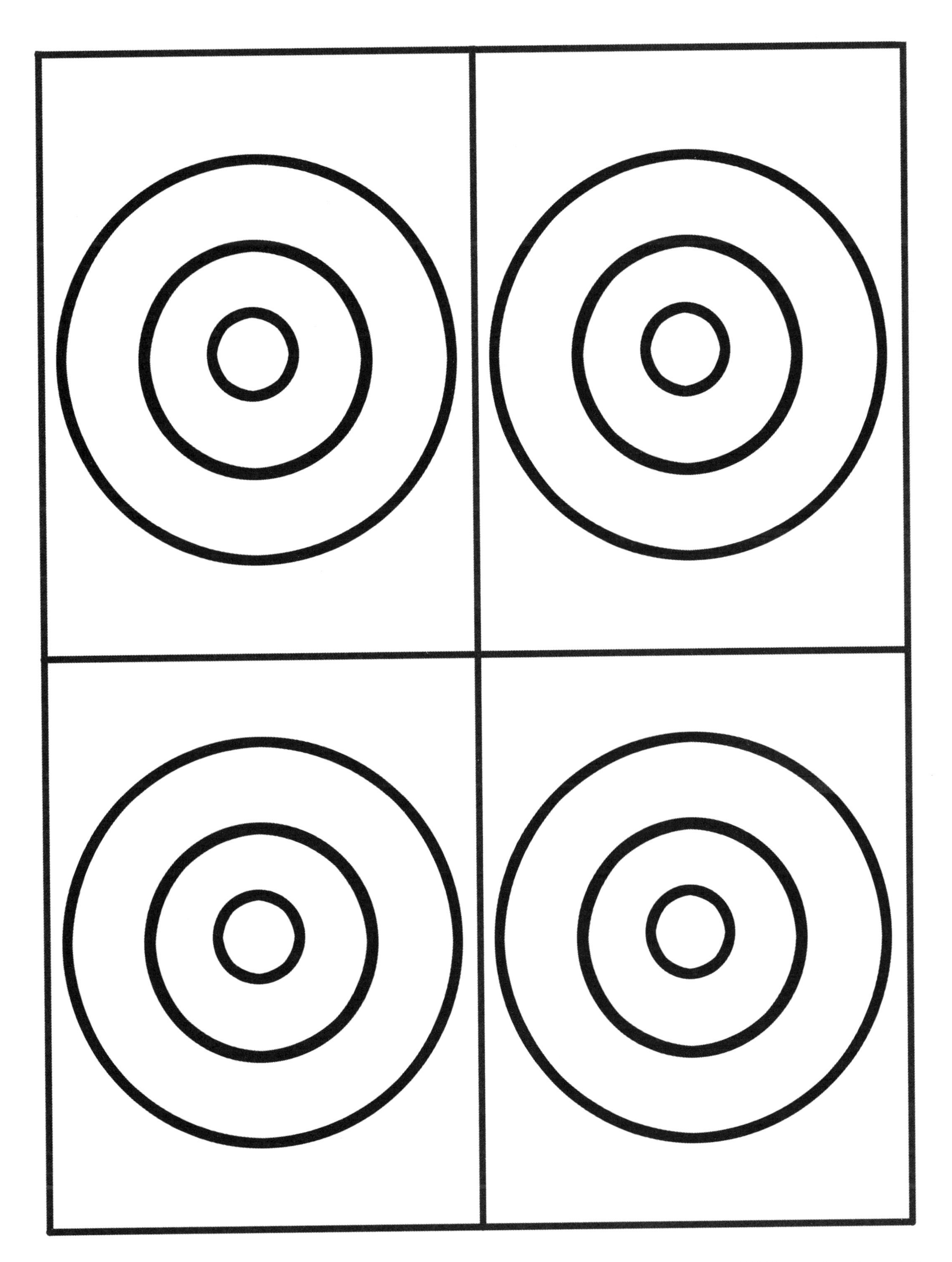

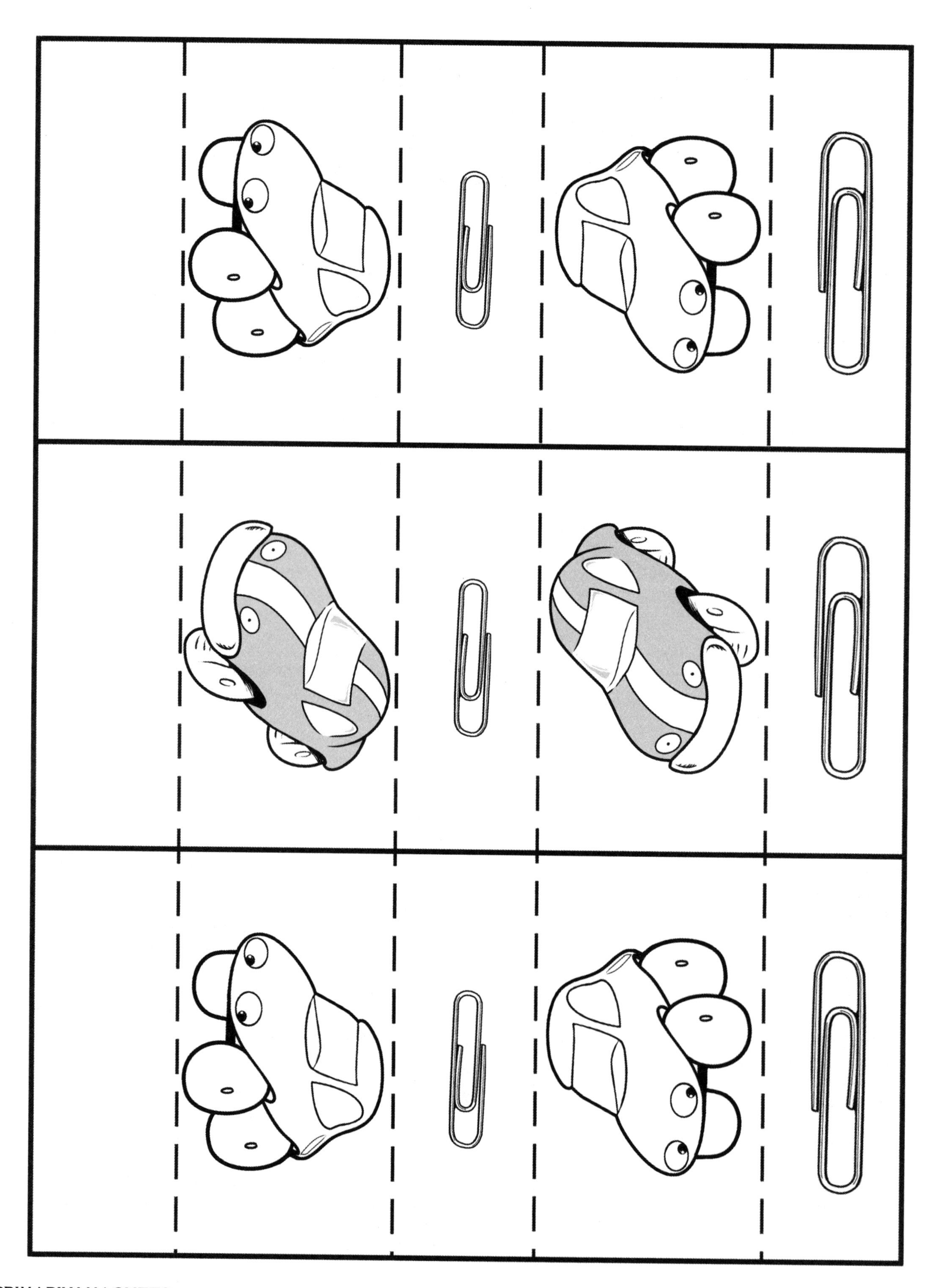

Topic
Magnets

Key Question
What is your explanation for each amazing feat?

Learning Goal
Students will how magnetic properties make each of the amazing feats work.

Guiding Documents
Project 2061 Benchmark
- *Without touching them, a magnet pulls on all things made of iron and either pushes or pulls on other magnets.*

NRC Standards
- *Magnets attract and repel each other and certain kinds of other materials.*
- *Employ simple equipment and tools to gather data and extend the senses.*

Science
Physical science
  force
    magnetism

Integrated Processes
Observing
Predicting
Comparing and contrasting
Applying

Materials
Ring magnets
Tape
File folder
Magnetic strip tape, 6 inches
Drinking straw
Paper clips
Cardboard strip, 30 cm x 10 cm
Thin cloth glove
Steel ball bearing
Ruler
Ballpoint pen or sharp pencil

Background Information
This investigation applies two basic characteristics of magnets, the interaction between two magnets (attraction/repulsion) and a magnet's interaction with magnetic objects through various materials.

Every magnet has a magnetic field that is strongest at its poles. The poles of a ring magnet are on its flat sides. If two magnets are placed facing each other with like poles together, they will repel each other; if unlike poles are together, they will attract.

Magnetic fields are able to pass through many materials. This can be demonstrated by placing a piece of paper between a magnet and a paper clip or between two magnets. The magnetic force can pass through the paper and attract the clip, or the two magnets can interact through the paper.

In this activity, several "amazing" things will happen before students' very eyes. Each event is possible because of magnetism. Students will observe the amazing feats and will be asked to explain how each works. For example, the frog is able to float because the two like magnetic poles are repelling each other; the bear is able to dance because the force of the magnetic strip passes through the cardstock and interacts with the magnetic force of the ring magnets.

Management
1. Magnetic strip tape can be found at craft stores.
2. Ring magnets (order number 1971) are available from AIMS.
3. Prior to presenting the amazing magnet feats, the following preparations are necessary:

The Dancing Bear
Gather two ring magnets and one piece of magnetic strip tape. Copy and color the stage scene. Attach it to a manila folder. Copy the dancing bear onto card stock. Color it and cut it out. Place a 3-inch piece of the magnetic strip tape in the center of the bear. Glue the front and back of the bear together.

The Floating Frog
Gather four ring magnets, tape, and a drinking straw. Copy the frog strip onto card stock. Color and cut it out. Wrap the strip around a stack of two magnets and attach with tape.

No Hands

Gather two magnets.

The Magic Touch

Place a 3-inch piece of magnetic strip tape inside the index finger of a thin cloth glove.

A Ball That Rolls Uphill

Gather a ruler, a ballpoint pen or sharp pencil, and a strip of cardboard (30 cm x 10 cm). Score the middle of the 30 cm length of cardboard to make it easier to fold it into a V-shaped valley. (To score the cardboard, position a ruler along the middle of the cardboard. Press hard with the pen or sharp pencil as you draw a line along the edge of the ruler. This breaks fibers in the cardboard, allowing for a crisp, straight fold.) Set one end of the cardboard on several books to make a ramp. You will also need a magnet and a steel ball bearing for this feat.

Procedure

1. Ask students about the types of amazing feats they might find at a circus. [Acrobatics, trapeze artists, balancing on a tightrope, defying gravity, etc.]
2. Tell them that just like circus feats, magnets can do amazing things. Ask what they know about magnets. [Magnets attract or repel each other and attract certain types of metal. They can interact through a variety of materials. Etc.]
3. Explain to the students that you are going to do a series of "amazing" feats and that you want them to observe closely because they will be asked to explain how each works. Present the following:

The Dancing Bear

Display the "stage" and bear suggested in the *Management* section of this activity. Secretly place the two magnets behind the folder so that they hold the dancing bear in place. Move the magnets to make the bear dance.

The Floating Frog

Place the prepared frog suggested in the *Management* section, onto a straw. Hide a stack of two additional magnets, arranged to repel the frog's magnet, in your fist and slide your fist up and down the straw making the frog float above your closed fist.

No Hands

Place a ring magnet on the inside and outside of one hand. Hold the open hand, palm side up, so that the class can see the magnet in your palm. Don't let them see the magnet on the back side of your hand. Explain that you will hold your hand straight up into the air and that the magnet will not fall; you will defy gravity.

The Magic Touch

Tell the class that you have an amazing glove. Wearing the glove suggested in the *Management* section of this activity, place your hand near a stack of paper clips and see how many you can pick up by touching them with your index finger. Then, place your gloved finger near other objects that would be attacked to a magnet and command the object to come to you.

A Ball That Rolls Uphill

Using the cardboard ramp described in *Management,* place a steel ball bearing near the top of the V. Release the ball. Place the ball bearing at the bottom of the hill and tell the class that you will defy gravity and make the ball roll up the cardboard hill. Hold a magnet behind the cardboard. Use the magnet to pull the ball up the hill.

4. End with a discussion about what magnetic properties allow each of the demonstrations to work. Allow students to try the various "feats" for themselves.

Connecting Learning

1. How can objects be moved? [pulling, pushing]
2. What is the pulling motion called when it involves magnets? [attraction] ...pushing motion? [repulsion]
3. How did each amazing feat work?
4. What can we do with a magnet?
5. Do like poles attract or repel each other?
6. Are magnets attracted to paper?
7. Name three things that are attracted to a magnet.

AMAZING
Feats With
Magnets

Tab

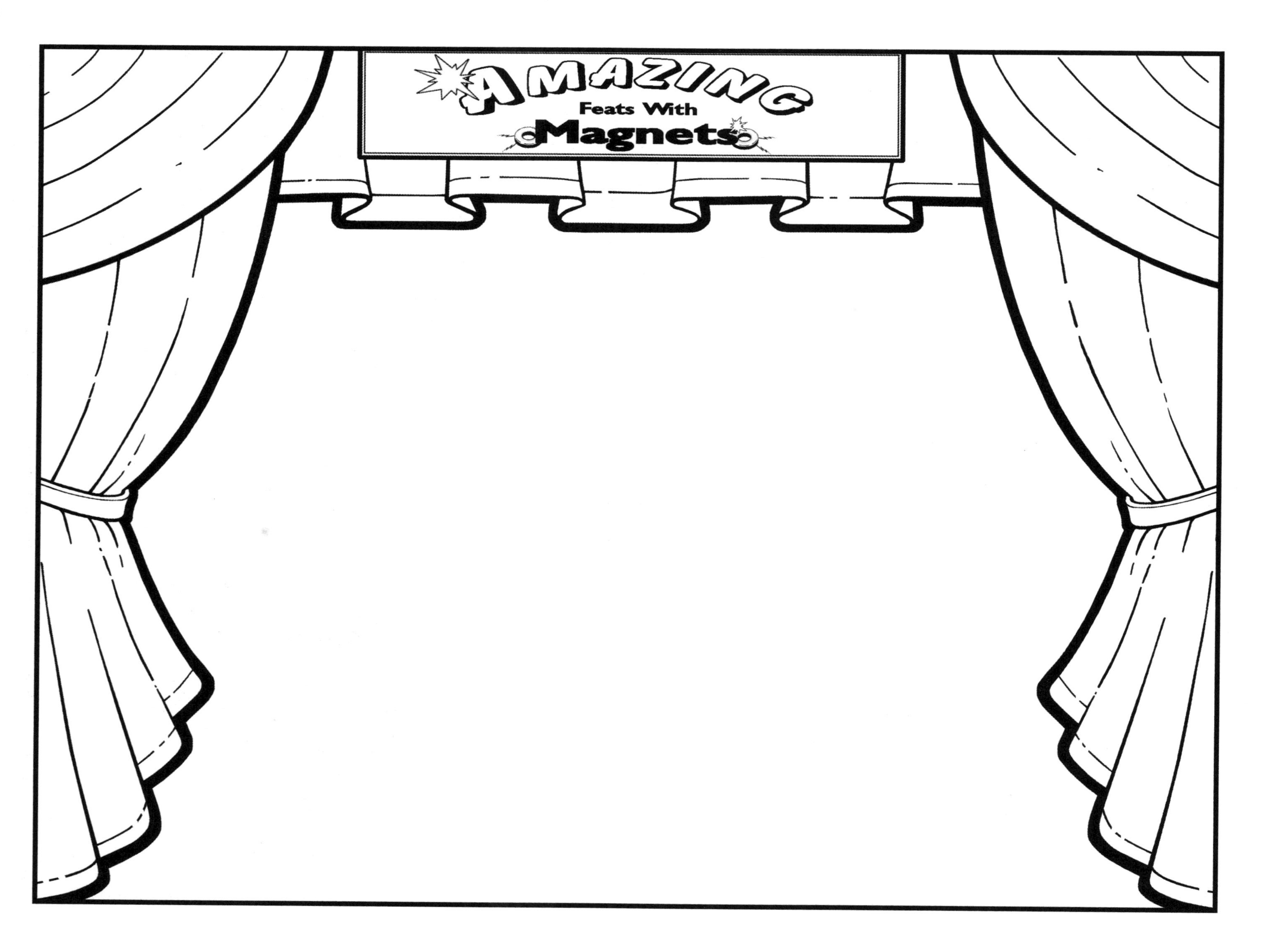
AMAZING
Feats With
Magnets

My Magnet Review
1.
Magnets can attract and repel other magnets.
Use pictures or words to explain this.
Attract
Repel
N
S
N
N
S
2.

Magnets can move
things without
touching them.
Magnets can work
through things.
3.
Magnets are used in
many ways.
Draw or list places
where they are used.
4.

Magnets will attract ...
5.
Magnets come
in many shapes
and sizes.
6.

Primarily Magnets

Assessment

1. A magnet will attract

A. 

B.

C. 

D. all of the above.

2. Which picture shows an object moving without being touched?

E. 

F. 

G. 

H. 

3. Which picture shows only things that a magnet would attract?

A. 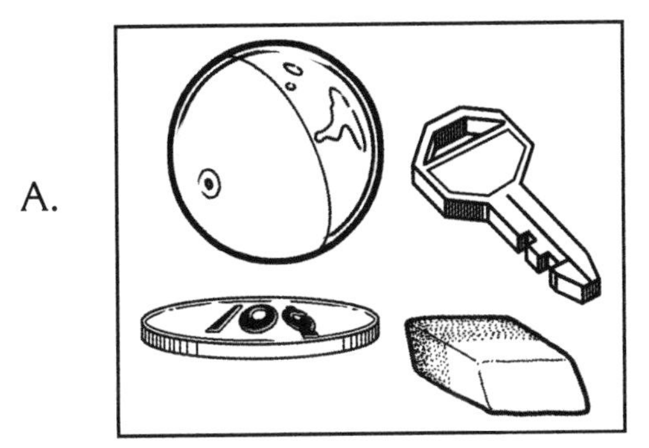

B. 

C. 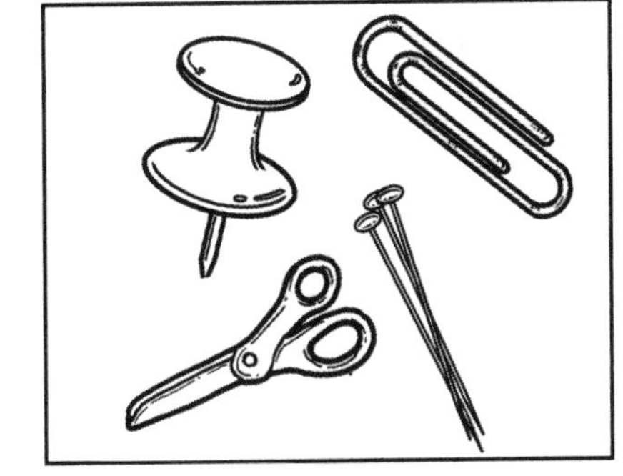

D. 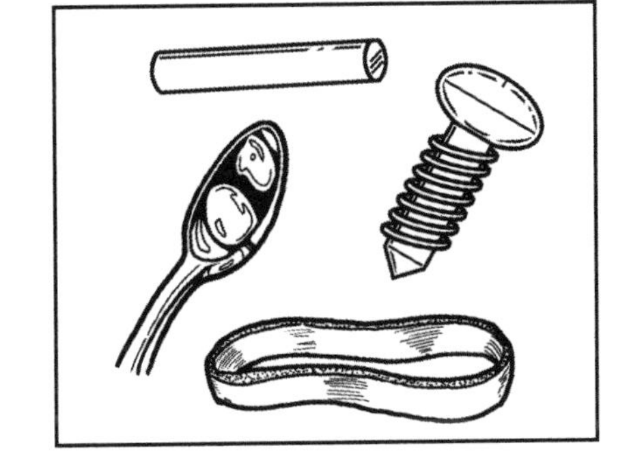

4. Look at the two magnets. If you push the magnets toward each other what will happen?

E. They will break into pieces.
F. They will be pulled together.
G. They will push apart.
H. They will turn in opposite directions.

5. Which of these pairs of magnets will repel each other?

A. 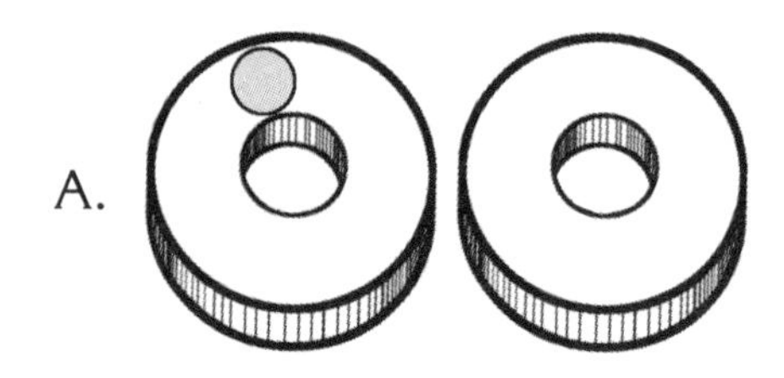

B. 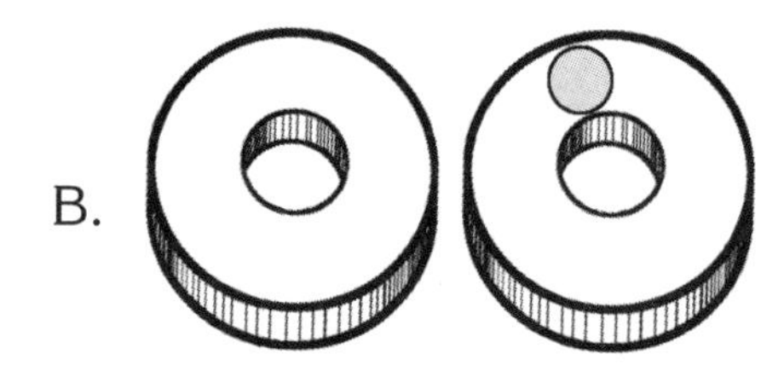

C. 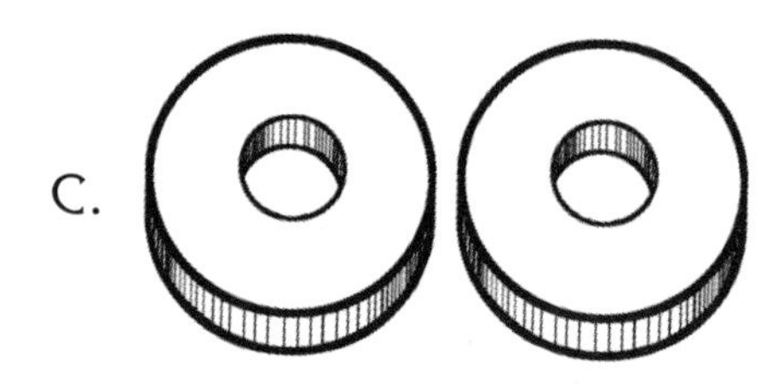

D. All of the above

6. What do you know about these two magnets?

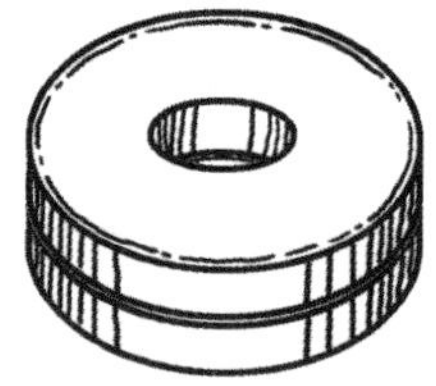

E. They are made of aluminum.
F. They are repelling each other.
G. They are attracted to each other.
H. They are purple.

7. A magnet can attract things through
A. cardboard
B. paper
C. plastic
D. all of the above.

8. Which of the following materials could block magnetic force?
E. A brick
F. A piece paper
G. A sheet of cardboard
H. A plastic bottle

9. Which of these everyday objects does not use a magnet?
A. Can opener
B. Credit card
C. Rake
D. Purse

10. Which of the following can be damaged by magnets?
E. Television
F. Credit cards
G. Video tapes
H. All of the above

Primarily Magnets

Assessment Key

1. C
2. F
3. C
4. F
5. C
6. G
7. D
8. E
9. C
10. H

Magnet Bibliography

Blevins, Wiley. *Magnets*. Compass Point Books. Mankato, MN. 2003. (A simple introduction to magnets, using and easy-to-read text that incorporates phonics instruction.)

Branley, Franklyn M. *What Makes a Magnet?* HarperTrophy. New York. 1996. (This books explores what things are attracted to a magnet. There is a simple explanation of how magnets work. It contains easy-to-do experiments for making a magnet and a compass.)

Branley, Franklyn. *Mickey's Magnet*. HarperTrophy. New York. 1976. (A young boy learns to pick up his mother's pins using a magnet. The story goes on to explore things that are magnetic and non-magnetic.)

Heinemann. *Magnets (My World of Science)*. Heinemann. Portsmouth, NH. 2005. (Learn what magnets are, where you can find them, how people use them in everyday life, and even how to make one.)

Kirkpatrick, Rena. *Magnets*. Raintree/Steck-Vaughn. Austin, TX. 1991. (*Look at Science Series*. Easy-to-read text and illustrations introduce facts about magnets.)

Krensky, Stephen. All About Magnets. Scholastic, Inc. New York. 1993. (A *Do-It-Yourself Science Book*. You will have fun doing the activities and games with the magnet included in the back of this book.)

Nelson, Robin. *Magnets*. Lerner Publications Co., Minneapolis. 2004. (This book has wonderful pictures and simple text of where magnets are used.)

Fowler, Allan. *What Can Magnets Do?* Sterling Publishing Co., Inc. New York. 1998. (Simple text and photographs to explain how magnets work. Includes uses of magnets.)

Rosinsky. Natalie M. Magnets: Pulling Together, Pushing Apart. Picture Window Books. Mankato, MN. 2004. (This fascinating book will attract readers while teaching them the fundamental facts about magnets.)

Schreiber, Anne. *Magnets*. Grosset & Dunlap. New York. 2003. (Delightfully illustrated, the book shows real-world uses of magnets. Contains interesting facts and easy-to-do magnet activities.)

Tocci, Salvatore. *Experiments With Magnets*. Children's Press. Chicago. 2002. (Book includes information on magnetic interactions, historical perspective on the use of magnetism, use of electromagnetism, etc.)

Walker, Sally M. *Magnetism*. Lerner Publications Co. Minneapolis, MN. 2006. (Chapter headings include: Magnets, Magnetic Materials, How Magnets Work, Magnetic Poles, Kinds of Magnets.)

The AIMS Program

AIMS is the acronym for "**A**ctivities **I**ntegrating **M**athematics and **S**cience." Such integration enriches learning and makes it meaningful and holistic. AIMS began as a project of Fresno Pacific University to integrate the study of mathematics and science in grades K-9, but has since expanded to include language arts, social studies, and other disciplines.

AIMS is a continuing program of the non-profit AIMS Education Foundation. It had its inception in a National Science Foundation funded program whose purpose was to explore the effectiveness of integrating mathematics and science. The project directors, in cooperation with 80 elementary classroom teachers, devoted two years to a thorough field-testing of the results and implications of integration.

The approach met with such positive results that the decision was made to launch a program to create instructional materials incorporating this concept. Despite the fact that thoughtful educators have long recommended an integrative approach, very little appropriate material was available in 1981 when the project began. A series of writing projects ensued, and today the AIMS Education Foundation is committed to continuing the creation of new integrated activities on a permanent basis.

The AIMS program is funded through the sale of books, products, and professional-development workshops, and through proceeds from the Foundation's endowment. All net income from programs and products flows into a trust fund administered by the AIMS Education Foundation. Use of these funds is restricted to support of research, development, and publication of new materials. Writers donate all their rights to the Foundation to support its ongoing program. No royalties are paid to the writers.

The rationale for integration lies in the fact that science, mathematics, language arts, social studies, etc., are integrally interwoven in the real world, from which it follows that they should be similarly treated in the classroom where students are being prepared to live in that world. Teachers who use the AIMS program give enthusiastic endorsement to the effectiveness of this approach.

Science encompasses the art of questioning, investigating, hypothesizing, discovering, and communicating. Mathematics is a language that provides clarity, objectivity, and understanding. The language arts provide us with powerful tools of communication. Many of the major contemporary societal issues stem from advancements in science and must be studied in the context of the social sciences. Therefore, it is timely that all of us take seriously a more holistic method of educating our students. This goal motivates all who are associated with the AIMS Program. We invite you to join us in this effort.

Meaningful integration of knowledge is a major recommendation coming from the nation's professional science and mathematics associations. The American Association for the Advancement of Science in *Science for All Americans* strongly recommends the integration of mathematics, science, and technology. The National Council of Teachers of Mathematics places strong emphasis on applications of mathematics found in science investigations. AIMS is fully aligned with these recommendations.

Extensive field testing of AIMS investigations confirms these beneficial results:

1. Mathematics becomes more meaningful, hence more useful, when it is applied to situations that interest students.
2. The extent to which science is studied and understood is increased when mathematics and science are integrated.
3. There is improved quality of learning and retention, supporting the thesis that learning which is meaningful and relevant is more effective.
4. Motivation and involvement are increased dramatically as students investigate real-world situations and participate actively in the process.

We invite you to become part of this classroom teacher movement by using an integrated approach to learning and sharing any suggestions you may have. The AIMS Program welcomes you!

AIMS Education Foundation Programs

When you host an AIMS workshop for elementary and middle school educators, you will know your teachers are receiving effective, usable training they can apply in their classrooms immediately.

AIMS Workshops are Designed for Teachers

- Correlated to your state standards;
- Address key topic areas, including math content, science content, and process skills;
- Provide practice of activity-based teaching;
- Address classroom management issues and higher-order thinking skills;
- Give you AIMS resources; and
- Offer optional college (graduate-level) credits for many courses.

AIMS Workshops Fit District/Administrative Needs

- Flexible scheduling and grade-span options;
- Customized (one-, two-, or three-day) workshops meet specific schedule, topic, state standards, and grade-span needs;
- Prepackaged four-day workshops for in-depth math and science training available (includes all materials and expenses);
- Sustained staff development is available for which workshops can be scheduled throughout the school year;
- Eligible for funding under the Title I and Title II sections of No Child Left Behind; and
- Affordable professional development—consecutive-day workshops offer considerable savings.

University Credit—Correspondence Courses

AIMS offers correspondence courses through a partnership with Fresno Pacific University.

- Convenient distance-learning courses—you study at your own pace and schedule. No computer or Internet access required!

Introducing AIMS State-Specific Science Curriculum

Developed to meet 100% of your state's standards, AIMS' State-Specific Science Curriculum gives students the opportunity to build content knowledge, thinking skills, and fundamental science processes.

- Each grade-specific module has been developed to extend the AIMS approach to full-year science programs. Modules can be used as a complete curriculum or as a supplement to existing materials.
- Each standards-based module includes math reading, hands-on investigations, and assessments.

Like all AIMS resources, these modules are able to serve students at all stages of readiness, making these a great value across the grades served in your school.

For current information regarding the programs described above, please complete the following form and mail it to: P.O. Box 8120, Fresno, CA 93747.

Information Request

Please send current information on the items checked:

____ *Basic Information Packet* on AIMS materials

____ Hosting information for AIMS workshops

____ AIMS State-Specific Science Curriculum

Name: ______________________________

Phone: ______________________ E-mail: ______________________

Address: ______________________________
Street City State Zip

AIMS Program Publications

Actions with Fractions, 4-9
The Amazing Circle, 4-9
Awesome Addition and Super Subtraction, 2-3
Bats Incredible! 2-4
Brick Layers II, 4-9
The Budding Botanist, 3-6
Chemistry Matters, 4-7
Counting on Coins, K-2
Cycles of Knowing and Growing, 1-3
Crazy about Cotton, 3-7
Critters, 2-5
Earth Book, 6-9
Electrical Connections, 4-9
Exploring Environments, K-6
Fabulous Fractions, 3-6
Fall into Math and Science, K-1
Field Detectives, 3-6
Finding Your Bearings, 4-9
Floaters and Sinkers, 5-9
From Head to Toe, 5-9
Fun with Foods, 5-9
Glide into Winter with Math and Science, K-1
Gravity Rules! 5-12
Hardhatting in a Geo-World, 3-5
It's About Time, K-2
It Must Be A Bird, Pre-K-2
Jaw Breakers and Heart Thumpers, 3-5
Looking at Geometry, 6-9
Looking at Lines, 6-9
Machine Shop, 5-9
Magnificent Microworld Adventures, 5-9
Marvelous Multiplication and Dazzling Division, 4-5
Math + Science, A Solution, 5-9
Mostly Magnets, 3-6
Movie Math Mania, 6-9
Multiplication the Algebra Way, 6-8
Off the Wall Science, 3-9
Out of This World, 4-8
Paper Square Geometry:
 The Mathematics of Origami, 5-12
Puzzle Play, 4-8
Pieces and Patterns, 5-9
Popping With Power, 3-5
Positive vs. Negative, 6-9
Primarily Bears, K-6
Primarily Earth, K-3
Primarily Physics, K-3
Primarily Plants, K-3
Problem Solving: Just for the Fun of It! 4-9
Problem Solving: Just for the Fun of It! Book Two, 4-9
Proportional Reasoning, 6-9
Ray's Reflections, 4-8
Sensational Springtime, K-2
Sense-Able Science, K-1
The Sky's the Limit, 5-9
Soap Films and Bubbles, 4-9
Solve It! K-1: Problem-Solving Strategies, K-1
Solve It! 2nd: Problem-Solving Strategies, 2
Solve It! 3rd: Problem-Solving Strategies, 3
Solve It! 4th: Problem-Solving Strategies, 4
Solve It! 5th: Problem-Solving Strategies, 5
Solving Equations: A Conceptual Approach, 6-9
Spatial Visualization, 4-9
Spills and Ripples, 5-12
Spring into Math and Science, K-1
Statistics and Probability, 6-9
Through the Eyes of the Explorers, 5-9
Under Construction, K-2
Water Precious Water, 2-6
Weather Sense: Temperature, Air Pressure, and Wind, 4-5
Weather Sense: Moisture, 4-5
Winter Wonders, K-2

Spanish Supplements*
Fall Into Math and Science, K-1
Glide Into Winter with Math and Science, K-1
Mostly Magnets, 2-8
Pieces and Patterns, 5-9
Primarily Bears, K-6
Primarily Physics, K-3
Sense-Able Science, K-1
Spring Into Math and Science, K-1

* Spanish supplements are only available as downloads from the AIMS website. The supplements contain only the student pages in Spanish; you will need the English version of the book for the teacher's text.

Spanish Edition
Constructores II: Ingeniería Creativa Con Construcciones LEGO® 4-9
The entire book is written in Spanish. English pages not included.

Other Publications
Historical Connections in Mathematics, Vol. I, 5-9
Historical Connections in Mathematics, Vol. II, 5-9
Historical Connections in Mathematics, Vol. III, 5-9
Mathematicians are People, Too
Mathematicians are People, Too, Vol. II
What's Next, Volume 1, 4-12
What's Next, Volume 2, 4-12
What's Next, Volume 3, 4-12

For further information, contact:
AIMS Education Foundation • P.O. Box 8120 • Fresno, California 93747-8120
www.aimsedu.org • 559.255.6396 (fax) • 888.733.2467 (toll free)

Duplication Rights

Standard Duplication Rights

Purchasers of AIMS activities (individually or in books and magazines) may make up to 200 copies of any portion of the purchased activities, provided these copies will be used for educational purposes and only at one school site.

Workshop or conference presenters may make one copy of a purchased activity for each participant, with a limit of five activities per workshop or conference session.

Standard duplication rights apply to activities received at workshops, free sample activities provided by AIMS, and activities received by conference participants.

All copies must bear the AIMS Education Foundation copyright information.

Unlimited Duplication Rights

To ensure compliance with copyright regulations, AIMS users may upgrade from standard to unlimited duplication rights. Such rights permit unlimited duplication of purchased activities (including revisions) for use at a given school site.

Activities received at workshops are eligible for upgrade from standard to unlimited duplication rights.

Free sample activities and activities received as a conference participant are not eligible for upgrade from standard to unlimited duplication rights.

Upgrade Fees

The fees for upgrading from standard to unlimited duplication rights are:
- $5 per activity per site,
- $25 per book per site, and
- $10 per magazine issue per site.

The cost of upgrading is shown in the following examples:
- activity: 5 activities x 5 sites x $5 = $125
- book: 10 books x 5 sites x $25 = $1250
- magazine issue: 1 issue x 5 sites x $10 = $50

Purchasing Unlimited Duplication Rights

To purchase unlimited duplication rights, please provide us the following:
1. The name of the individual responsible for coordinating the purchase of duplication rights.
2. The title of each book, activity, and magazine issue to be covered.
3. The number of school sites and name of each site for which rights are being purchased.
4. Payment (check, purchase order, credit card)

Requested duplication rights are automatically authorized with payment. The individual responsible for coordinating the purchase of duplication rights will be sent a certificate verifying the purchase.

Internet Use

Permission to make AIMS activities available on the Internet is determined on a case-by-case basis.

P. O. Box 8120, Fresno, CA 93747-8120
aimsed@aimsedu.org • www.aimsedu.org
559.255.6396 (fax) • 888.733.2467 (toll free)

